NOT BORN YESTERDAY

NOT BORN YESTERDAY

THE SCIENCE OF *Who* WE TRUST
AND *What* WE BELIEVE

HUGO MERCIER

PRINCETON UNIVERSITY PRESS
PRINCETON & OXFORD

Copyright © 2020 by Hugo Mercier

Requests for permission to reproduce material from this work should be sent to permissions@press.princeton.edu

Published by Princeton University Press

41 William Street, Princeton, New Jersey 08540

6 Oxford Street, Woodstock, Oxfordshire OX20 1TR

press.princeton.edu

All Rights Reserved

ISBN 978-0-691-17870-7

ISBN (e-book) 978-0-691-19884-2

British Library Cataloging-in-Publication Data is available

Editorial: Sarah Caro, Charlie Allen, and Hannah Paul

Production Editorial: Jenny Wolkowicki

Text design: Leslie Flis

Jacket design: Michel Vrana

Production: Erin Suydam

Publicity: Maria Whelan and Kate Farquhar-Thomson

Jacket art: iStock

This book has been composed in Arno Pro and Heading Smallcase Pro

Printed on acid-free paper. ∞

Printed and bound in Great Britain by Clays Ltd, Elcograf S.p.A.

10 9 8 7 6 5 4 3 2 1

To Thérèse Cronin

CONTENTS

List of Illustrations ix
Acknowledgments xi
Introduction xiii

1 The Case for Gullibility — 1
2 Vigilance in Communication — 15
3 Evolving Open-Mindedness — 30
4 What to Believe? — 47
5 Who Knows Best? — 63
6 Who to Trust? — 78
7 What to Feel? — 95
8 Demagogues, Prophets, and Preachers — 113
9 Propagandists, Campaigners, and Advertisers — 128
10 Titillating Rumors — 146

11	From Circular Reporting to Supernatural Beliefs	166
12	Witches' Confessions and Other Useful Absurdities	181
13	Futile Fake News	199
14	Shallow Gurus	217
15	Angry Pundits and Skillful Con Men	240
16	The Case against Gullibility	257

Notes 273

References 307

Index 351

LIST OF ILLUSTRATIONS

FIGURE 1. The lines in the Asch conformity experiments.   6

FIGURE 2. "Bridge" strip from the webcomic *xkcd* by Randall Munroe.   72

FIGURE 3. Two examples of pareidolia: seeing faces where there are none.   157

FIGURE 4. What path does a ball launched at the arrow follow when it exits the tube?   224

ACKNOWLEDGMENTS

THE IDEA FOR THIS BOOK stems from the article "Epistemic Vigilance," written by Dan Sperber, Fabrice Clément, Christophe Heintz, Olivier Mascaro, Gloria Origgi, Deirdre Wilson, and myself. In this article we suggested that humans are endowed with cognitive mechanisms dedicated to evaluating communicated information. I am particularly grateful to Dan Sperber—thesis supervisor, coauthor, mentor, friend. Besides having shaped my ideas through his writings and his discussions, he patiently read and gave me feedback on the book. Fabrice Clément had written his PhD thesis and a book—*Les Mécanismes de la crédulité* (The mechanisms of credulity)—on the same theme, and we discussed these issues when I was a postdoctoral researcher at the University of Neuchâtel. Besides Dan and Fabrice, the ideas in this book have been shaped by the feedback from the students of the Communication, Trust, and Argumentation class from 2018 to 2019, and by discussions at the Department of Cognitive Studies of the ENS in Paris; the University of Pennsylvania; and the countless conferences, restaurants, pubs, and cafés where I've badgered people with the idea that humans aren't gullible. Lila San Roque generously shared fantastic examples of evidential use among the Duna, and I benefited from Chris Street's knowledge of the lie detection literature.

My deepest thanks go to those who have commented on the whole, or parts of the manuscript: Sacha Altay (twice!), Stefaan Blancke, Pascal Boyer, Coralie Chevallier, Thérèse Cronin (twice as well!), Guillaume Dezecache, Helena Miton, Olivier Morin, Thom Scott-Phillips, Dan Sperber, and Radu Umbres.

This book would not have existed without my agents John and Max Brockman, Sarah Caro, the editor who believed in the project from the start and provided very valuable feedback, as well as the team at Princeton University Press.

I have benefited from the financial backing of the Direction Générale de l'Armement (thanks to Didier Bazalgette in particular); the Philosophy, Politics, and Economics program at the University of Pennsylvania (with the generous support of Steven F. Goldstone); the University of Neuchâtel's Cognitive Science group and the Swiss National Science Foundation (Ambizione grant no. PZ00P1_142388); the Agence Nationale de la Recherche (grant EUR FrontCog ANR-17-EURE-0017 to the DEC, and grant ANR-16-TERC-0001-01 to myself); and last but not least, the Centre National de la Recherche Scientifique, my current employer, allowing me to work in the amazing place that is the Jean Nicod Institute. In particular, the team I belong to—Evolution and Social Cognition, composed of Jean-Baptiste André, Nicolas Baumard, Coralie Chevallier, Olivier Morin, and our students, engineers, and postdocs—provides the best social and intellectual environment one could hope for.

I will never stop thanking my parents, grandparents, and extended family for their unwavering support. Christopher and Arthur are the best boys in the world; they have taught me much about love. They have also taught me that children aren't gullible, even when we wish they'd be a bit easier to influence. Thérèse's encouragement has meant more to me than I can ever communicate. Thank you for everything.

INTRODUCTION

AS I WAS WALKING BACK from university one day, a respectable-looking middle-aged man accosted me. He spun a good story: he was a doctor working in the local hospital, he had to rush to some urgent doctorly thing, but he'd lost his wallet, and he had no money for a cab ride. He was in dire need of twenty euros. He gave me his business card, told me I could call the number and his secretary would wire the money back to me shortly.

After some more cajoling I gave him twenty euros.

There was no doctor of this name, and no secretary at the end of the line.

How stupid was I?

And how ironic that, twenty years later, I would be writing a book arguing that people aren't gullible.

THE CASE FOR GULLIBILITY

If you think I'm gullible, wait until you meet, in the pages that follow, people who believe that the earth is a flat disk surrounded by a two-hundred-foot wall of ice, "*Game of Thrones*–style,"[1] that witches poison their cattle with magical darts, that the local Jews kill young boys to drink their blood as a Passover ritual, that high-up Democratic operatives oversee a pedophile ring out of a pizza joint, that former North Korean leader Kim Jong-il could teleport and control the weather, or that former U.S. president Barack Obama is a devout Muslim.

Look at all the gibberish transmitted through TV, books, radio, pamphlets, and social media that ends up being accepted by large swaths of the population. How could I possibly be claiming that we aren't gullible, that we don't accept whatever we read or hear?

Arguing against widespread credulity puts me in the minority. A long line of scholarship—from ancient Greece to twenty-first-century America, from the most progressive to the most reactionary—portrays the mass of people as hopelessly gullible. For most of history, thinkers have based their grim conclusions on what they thought they observed: voters submissively following demagogues, crowds worked up into rampages by bloodthirsty leaders, masses cowing to charismatic personalities. In the mid-twentieth century, psychological experiments brought more grist to this mill, showing participants blindly obeying authority, believing a group over the clear evidence of their own eyes. In the past few decades, a series of sophisticated models have appeared that provide an explanation for human gullibility. Here is the core of their argument: we have so much to learn from others, and the task of figuring out who to learn from is so difficult, that we rely on simple heuristics such as "follow the majority" or "follow prestigious individuals." Humans would owe their success as a species to their capacity to absorb their local culture, even if that means accepting some maladaptive practices or mistaken beliefs along the way.

The goal of this book is to show this is all wrong. We don't credulously accept whatever we're told—even if those views are supported by the majority of the population, or by prestigious, charismatic individuals. On the contrary, we are skilled at figuring out who to trust and what to believe, and, if anything, we're too hard rather than too easy to influence.

THE CASE AGAINST GULLIBILITY

Even if suggestibility might have some advantages in helping us acquire skills and beliefs from our cultural environment, it is simply too costly to be a stable, persistent state of affairs, as I will argue in chapter 2. Accepting whatever others are communicating only pays off if their interests are aligned with ours—think cells in a body, bees in a beehive. As far as communication between humans is concerned, such commonality of interests is rarely achieved; even a pregnant mother has reasons to mistrust the chemical signals sent by her fetus. Fortunately, there are ways of making communication work even in the most adversarial of relationships. A prey can convince a predator not to chase it. But for such communication to occur, there must be strong guarantees that those who receive the signal will be better off believing it. The messages have to be kept, on the whole, honest. In the case of humans, honesty is maintained by a set of cognitive mechanisms that evaluate communicated information. These mechanisms allow us to accept most beneficial messages—to be open—while rejecting most harmful messages—to be vigilant. As a result, I have called them *open vigilance mechanisms*, and they are at the heart of this book.[2]

What about the "observations" used by so many scholars to make the case for gullibility? Most are merely popular misconceptions. As the research reviewed in chapters 8 and 9 shows, those who attempt to persuade the masses—from demagogues to advertisers, from preachers to campaign operatives—nearly always fail miserably. Medieval peasants in Europe drove many a priest to despair with their stubborn resistance to Christian precepts. The net effect on presidential elections of sending flyers, robocalling, and other campaign tricks is close to zero. The supposedly all-powerful Nazi propaganda machine barely affected its audience—it couldn't even get the Germans to like the Nazis.

Sheer gullibility predicts that influence is easy. It is not. Still, indubitably, people sometimes end up professing the most absurd views. What we must explain are the patterns: why some ideas, including good ones, are so hard to get across, while others, including bad ones, are so popular.

MECHANISMS OF OPEN VIGILANCE

Understanding our mechanisms of open vigilance is the key to making sense of the successes and failures of communication. These mechanisms process a variety of cues to tell us how much we should believe what we're told. Some mechanisms examine whether a message is compatible with what we already believe to be true, and whether it is supported by good arguments. Other mechanisms pay attention to the source of the message: Is the speaker likely to have reliable information? Does she have my interests at heart? Can I hold her accountable if she proves mistaken?

I review a wealth of evidence from experimental psychology showing how well our mechanisms of open vigilance function, including in small children and babies. It is thanks to these mechanisms that we reject most harmful claims. But these mechanisms also explain why we accept a few mistaken ideas.

For all their sophistication, and their capacity to learn and incorporate novel information, our mechanisms of open vigilance are not infinitely malleable. You, dear reader, are in an information environment that differs in myriad ways from the one your ancestors evolved in. You are interested in people you'll never meet (politicians, celebrities), events that don't affect you (a disaster in a distant country, the latest scientific breakthrough), and places you'll never visit (the bottom of the ocean, galaxies far, far away). You receive much information with no idea of where it came from: Who started the rumor that Elvis wasn't dead? What is the source of your parents' religious beliefs? You are asked to pass judgment on views that had no practical relevance whatsoever for our ancestors: What is the shape of the earth? How did life evolve? What is the best way to organize a large economic system? It would be surprising indeed if our mechanisms of open vigilance functioned impeccably in this brave new, and decidedly bizarre, world.

Our current informational environment pushes open vigilance mechanisms outside of their comfort zone, leading to mistakes. On the whole, we are more likely to reject valuable messages—from the reality of climate change to the efficacy of vaccination—than to accept inaccurate ones. The main exceptions to this pattern stem not so much from a failure of open vigilance itself, but from issues with the material it draws on. People sensibly use their own knowledge, beliefs, and intuitions to evaluate what they're told. Unfortunately, in some domains our intuitions appear to be quite systematically mistaken. If you had nothing else to go on, and someone told you that you were standing on a flat surface (rather than, say, a globe), you would spontaneously believe them. If you had nothing else to go on, and someone told you all your ancestors had always looked pretty much like you (and not like, say, fish), you would spontaneously believe them. Many popular yet mistaken beliefs spread not because they are pushed by masters of persuasion but because they are fundamentally intuitive.

If the flatness of the earth is intuitive, a two-hundred-foot-high, thousands-of-miles-long wall of ice is not. Nor is, say, Kim Jong-il's ability to teleport. Reassuringly, the most out-there beliefs out there are accepted only nominally. I bet a flat-earther would be shocked to actually run into that two-hundred-foot wall of ice at the end of the ocean. Seeing Kim Jong-il being beamed *Star Trek*–style would have confused the hell out of the dictator's most groveling sycophant. The critical question for understanding why such beliefs spread is not why people accept them, but why people profess them. Besides wanting to share what we take to be accurate views, there are many reasons for professing beliefs: to impress, annoy, please, seduce, manipulate, reassure. These goals are sometimes best served by making statements whose relation to reality is less than straightforward—or even, in some cases, statements diametrically opposed to the truth. In the face of such motivations, open vigilance mechanisms come to be used, perversely, to identify not the most plausible but the most implausible views.

From the most intuitive to the most preposterous, if we want to understand why some mistaken views catch on, we must understand how open vigilance works.

UPTAKE

At the end of the book, you should have a grasp on how you decide what to believe and who to trust. You should know more about how miserably unsuccessful most attempts at mass persuasion are, from the most banal—advertising, proselytizing—to the most extreme—brainwashing, subliminal influence. You should have some clues about why (some) mistaken ideas manage to spread, while (some) valuable insights prove so difficult to diffuse. You should understand why I once gave a fake doctor twenty euros.

I do hope you come to accept the core of the book's argument. But, please, don't just take my word for it. I'd hate to be proven wrong by my own readers.

1

THE CASE FOR GULLIBILITY

FOR MILLENNIA, people have accepted many bizarre beliefs and have been persuaded to engage in irrational behaviors (or so it appears). These beliefs and behaviors gave credence to the idea that the masses are gullible. In reality I believe the story is more complicated (or even completely different, as we'll see in the following chapters). But I must start by laying out the case for gullibility.

In 425 BCE, Athens had been locked for years in a mutually destructive war with Sparta. At the Battle of Pylos, the Athenian naval and ground forces managed to trap Spartan troops on the island of Sphacteria. Seeing that a significant number of their elite were among the captives, the Spartan leaders sued for peace, offering advantageous terms to Athens. The Athenians declined the offer. The war went on, Sparta regained the edge, and when a (temporary) peace treaty was signed, in 421 BCE, the terms were much less favorable to Athens. This blunder was only one of a series of terrible Athenian decisions. Some were morally repellent—killing all the citizens of a conquered city—others were strategically disastrous—launching a doomed expedition to Sicily. In the end, Athens lost the war and would never regain its former power.

In 1212, a "multitude of paupers" in France and Germany took the cross to fight the infidels and reclaim Jerusalem for the Catholic Church.[1] As many of these paupers were very young, this movement was dubbed the Children's Crusade. The youth made it to Saint-Denis, prayed in the cathedral, met the French king, hoped for a miracle. No miracle happened. What can be expected of an army of untrained, unfunded, disorganized preteens? Not much, which is what they achieved: none reached Jerusalem, and many died along the way.

In the mid-eighteenth century the Xhosa, a pastoralist people of South Africa, were suffering under the newly imposed British rule. Some of the Xhosa believed killing all their cattle and burning their crops would raise a ghost army that would fend off the British. They sacrificed thousands of heads of cattle and set fire to their fields. No ghost army arose. The British stayed. The Xhosa died.

On December 4, 2016, Edgar Maddison Welch entered the Comet Ping Pong pizzeria in Washington, DC, carrying an assault rifle, a revolver, and a shotgun. He wasn't there to rob the restaurant. Instead, he wanted to make sure that no children were being held hostage in the basement. There had been rumors that the Clintons—the former U.S. president and his wife, then campaigning for the presidency—were running a sex trafficking ring, and that Comet Ping Pong was one of their lairs. Welch was arrested and is now serving a prison sentence.

BLIND TRUST

Scholars, feeling superior to the masses, have often explained these questionable decisions and weird beliefs by a human disposition to be overly trusting, a disposition that would make the masses instinctively defer to charismatic leaders regardless of their competence or motivations, believe whatever they hear or read irrespective of its plausibility, and follow the crowd even when doing so leads to disaster. This explanation—the masses are credulous—has proven very influential throughout history even if, as will soon become clear, it is misguided.

Why did the Athenians lose the war against Sparta? Starting with Thucydides, chronicler of the Peloponnesian War, many commentators have blamed the influence of demagogues such as Cleon, a parvenu "very powerful with the multitude," who was deemed responsible for some of the war's worst blunders.[2] A generation later, Plato extended Thucydides's argument into a general indictment of democracy. For Plato, the rule of the many unavoidably gives rise to leaders who, "having a mob entirely at [their] disposal," turn into tyrants.[3]

Why would a bunch of youngsters abandon their homes in the vain hope of invading a faraway land? They were responding to the calls for a new crusade launched by Pope Innocent III, their supposed credulity inspiring the legend of the Pied Piper of Hamelin, whose magic flute grants him absolute power over all the children who hear it.[4] People's crusades also help explain the accusations that emerged in the Enlightenment, by the likes of the Baron d'Holbach, who chastised the Christian Church for "deliver[ing] mankind into [the] hands of [despots and tyrants] as a herd of slaves, of whom they may dispose at their pleasure."[5]

Why did the Xhosa kill their cattle? A century earlier, the Marquis de Condorcet, a central figure of the French Enlightenment, suggested that members of small-scale societies suffered from the "credulity of the first dupes," putting too much faith in "charlatans and sorcerers."[6] The Xhosa seem to fit this picture. They were taken in by Nongqawuse, a young prophetess who had had visions of the dead rising to fight the British, and of a new world in which "nobody would ever lead a troubled life. People would get whatever they wanted. Everything would be available in abundance."[7] Who would say no to that? Apparently not the Xhosa.

Why did Edgar Maddison Welch risk jail to deliver nonexistent children from the nonexistent basement of a harmless pizzeria? He had been listening to Alex Jones, the charismatic radio host who specializes in the craziest conspiracy theories, from the great Satanist takeover of America to government-sponsored calamities.[8] For a time, Jones took up the idea that the Clintons and their aides led an organization trafficking children for sex. As a *Washington Post* reporter put it, Jones and his ilk can peddle their wild theories because "gullibility helps create a market for it."[9]

All of these observers agree that people are often credulous, easily accept unsubstantiated arguments, and are routinely talked into stupid and costly behaviors. Indeed, it is difficult to find an idea that so well unites radically different thinkers. Preachers lambaste the "credulous multitude" who believe in gods other than the preachers' own.[10] Atheists point out "the almost superhuman gullibility" of those who follow religious preachers, whatever their god might be.[11] Conspiracy theorists feel superior to the "mind controlled sheeple" who accept the official news.[12] Debunkers think conspiracy theorists "super gullible" for believing the tall tales peddled by angry entertainers.[13] Conservative writers accuse the masses of criminal credulity when they revolt, prodded by shameless demagogues and driven mad by contagious emotions. Old-school leftists explain the passivity of the masses by their acceptance of the dominant ideology: "The individual lives his repression 'freely' as his own life: he desires what he is supposed to desire," instead of acting on "his original instinctual needs."[14]

THE CASE FOR GULLIBILITY

For most of history, the concept of widespread credulity has been fundamental to our understanding of society. The assumption that people are easily taken in by demagogues runs across Western thought, from ancient Greece to the Enlightenment, creating "political philosophy's central reason for skepticism about democracy."[15] Contemporary commenters still deplore how easily politicians sway voters by "pander[ing] to their gullibility."[16] But the ease with which people can be influenced has never been so (apparently) well illustrated as through a number of famous experiments conducted by social psychologists since the 1950s.

PSYCHOLOGISTS OF GULLIBILITY

First came Solomon Asch. In his most famous experiment he asked people to answer a simple question: Which of three lines (depicted in figure 1) is as long as the first line?[17] The three lines were clearly of different lengths, and one of them was an obvious match for the first. Yet participants made a mistake more than 30 percent of the time. Why would people provide such blatantly wrong answers? Before each participant was asked for their opinion, several participants had already replied. Unbeknownst to the actual participant, these other participants were confederates, planted by the experimenter. On some trials, all the confederates agreed on one of the wrong answers. These confederates held no power over the participants, who did not even know them, and they were providing plainly wrong answers. Still, more than 60 percent of participants chose at least once to follow the group's lead. A textbook written by Serge Moscovici, an influential social psychologist, describes these results as "one of the most dramatic illustrations of conformity, of blindly going along with the group, even

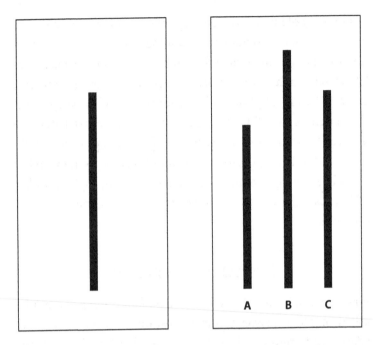

FIGURE 1. The lines in the Asch conformity experiments. *Source:* Wikipedia.

when the individual realizes that by doing so he turns his back on reality and truth."[18]

After Solomon Asch came Stanley Milgram. Milgram's first famous study was, like Asch's experiments, a study of conformity. He asked some of his students to stand on a sidewalk, looking at a building's window, and counted how many of the people passing by would imitate them.[19] When enough students were looking in the same direction—the critical group size seemed to be about five—nearly all those who passed by followed the students in looking at the building. It was as if people could not help but follow the crowd.

But Milgram is best known for a later, much more provocative experiment.[20] In this study, participants were asked to take

THE CASE FOR GULLIBILITY 7

part in research bearing ostensibly on learning. In the lab, they were introduced to another participant—who, once again, was actually a confederate. The experimenter pretended to randomly pick one of the two—always the confederate—to be the learner. Participants were then told the study tested whether someone who was motivated to avoid electric shocks would learn better. The learner had to memorize a list of words; when he made a mistake, the participant would be asked to administer an electric shock.

The participants sat in front of a big machine with a series of switches corresponding to electric shocks of increasingly high voltage. The confederate was led slightly away, to an experimental booth, but the participants could still hear him through a microphone. At first, the confederate did a good enough job memorizing the words, but as the task grew more difficult, he started making mistakes. The experimenter prompted the participants to shock the confederate, and all of them did. This was hardly surprising, as the first switches were marked as delivering only a "slight shock." As the confederate kept making mistakes, the experimenter urged the participants to increase the voltage. The switches went from "slight shock," to "moderate shock," then "strong shock," and "very strong shock," yet all the participants kept flipping the switches. It was only on the last switch of the "intense shock" series—300 volts—that a few participants refused to proceed. All the while, the confederate expressed his discomfort. At some point, he started howling in pain, begging the participants to stop: "Let me out of here! You can't hold me here! Get me out of here!"[21] He even complained of heart problems. Yet the vast majority of participants kept going.

When the "extreme intensity shock" series began, a few more participants stopped. One participant refused to go on when the switches indicated "danger: severe shock." At this stage, the confederate had simply stopped screaming and was begging to be freed. He then became completely unresponsive. But that didn't stop two-thirds of the participants from flipping the last two switches, 435 volts and 450 volts, marked with an ominous "XXX." Milgram had gotten a substantial majority of these ordinary American citizens to deliver (what they thought to be) potentially lethal electric shocks to a fellow citizen who (they thought) was writhing in pain and begging for mercy.

When learning of these results, and of a litany of historical cases seemingly attesting to similar phenomena, it is hard not to agree with the sweeping indictment leveled by political philosopher Jason Brennan: "Human beings are wired not to seek truth and justice but to seek consensus. They are shackled by social pressure. They are overly deferential to authority. They cower before uniform opinion. They are swayed not so much by reason but by a desire to belong, by emotional appeal, and by sex appeal."[22] Psychologist Daniel Gilbert and his colleagues concur: "That human beings are, in fact, more gullible than they are suspicious should probably 'be counted among the first and most common notions that are innate in us.'"[23]

If you believe that humans are by nature credulous, the natural question to ask is: Why? Already in 500 BCE Heraclitus, one of the first recorded Greek philosophers, was wondering:

> What use are the people's wits
> who let themselves be led
> by speechmakers, in crowds,
> without considering
> how many fools and thieves
> they are among, and how few
> choose the good?[24]

Heraclitus was echoed twenty-five hundred years later in a less poetic but more concise manner by this headline from the BBC: "Why are people so incredibly gullible?"[25]

ADAPTIVE CREDULITY

If social psychologists seem to have been bent on demonstrating human credulity, anthropologists have, for the most part, taken it for granted.[26] Many have seen the persistence of traditional beliefs and behaviors as unproblematic: children simply imbibe the culture that surrounds them, thereby ensuring its continuity. Logically, anthropologists have devoted little attention to children, who are supposed to be mere receptacles for the knowledge and skills of the previous generation.[27] Critical anthropologists have described the assumption that people absorb whatever culture surrounds them as the theory of "exhaustive cultural transmission,"[28] or, more pejoratively, as the "'fax model' of internalization."[29]

For all its simplicity, this model of cultural transmission helps us understand why people would be credulous: so they learn the knowledge and skills acquired by generations of their ancestors. Biologist Richard Dawkins thus explains the "programmed-in gullibility of a child" by its "useful[ness] for learning language and traditional wisdom."[30]

While it is easy to think of "traditional wisdom" one would rather not inherit from one's elders, from the belief in witchcraft to the practice of foot binding, these harmful customs are the exception. On the whole, most culturally acquired beliefs are sensible enough. Every day, we engage in culturally influenced behaviors too numerous to count: being able to speak, for a start, but also brushing our teeth, getting dressed, cooking, shopping, and so on.

Archaeological and anthropological evidence also suggests that cultural skills have been crucial to human survival for a very long time. Members of small-scale societies rely on traditional knowledge and know-how for foraging, hunting, processing food, making clothing, and producing the variety of tools indispensable to their survival.[31]

If the simplicity of this "fax model" of cultural transmission highlights the many benefits of learning from one's surrounding culture, its limits are also obvious. For one thing, it vastly underestimates the degree of cultural variation present even in the smallest, most self-contained societies. If some behaviors might be performed by all group members in a very similar fashion— some ritual, say—most activities exhibit significant variation. Not every hunter draws the same lessons from a set of tracks. Not every forager has the same techniques for finding berries. Not every artist creates equally appealing songs or sculptures or drawings. So even an individual bent on blindly copying the previous generation must make decisions: Who to copy from?

One of the most advanced frameworks for answering this question has been created by an anthropologist, Robert Boyd, and a biologist, Peter Richerson.[32] Known as *gene-culture coevolution*, this theory suggests that genes and cultures have influenced each other in the course of human evolution. In particular, Boyd and Richerson claim that culture has shaped our biological evolution. If choosing which bits of one's culture to copy is so important, then we should have evolved, through natural selection, mechanisms that help solve this problem as effectively as possible. We already have evolved dispositions that tackle a variety of issues our ancestors faced: forming a broadly accurate representation of our surroundings, picking edible food, avoiding predators, attracting mates, forming friendships, and so forth.[33] It would make sense that we had also evolved mechanisms to help us acquire the culture of our peers and our elders.

To solve the problem of who to learn from, we can start by looking at who performs well. Alex is an excellent cook; Renée is great at maintaining good social relationships; it makes sense to learn from them. But even when we have narrowed down the problem in this way, we're left with many potential actions to imitate. How do we work out exactly how and why Alex was able to cook such a great dish? Our intuitions help us rule out some factors—it probably wasn't his hairdo—but there remain many possibilities, ranging from the most obvious, such as the ingredients or the cooking time, to the least, such as the specific type of onions used or how the rice was stirred. As we find out when we try replicating a cook's recipe, the determinants of success can sometimes be quite opaque.[34]

To help us learn better from others, Boyd, Richerson, and their colleagues—such as anthropologist Joe Henrich or biologist Kevin Laland—suggest that humans are endowed with a series of rough heuristics to guide their cultural learning.[35] One of these rules of thumb extends our ability to learn from the most successful. Because it can be difficult to tell which of a successful individual's actions are responsible for their success—why Alex was able to produce a given dish well, say—it might be safer to copy indiscriminately everything successful people do and think, down to their appearance or hairdo. We can call this a *success bias*.

Another heuristic consists in copying whatever the majority does—the *conformity bias*.[36] This bias makes sense under the reasonable assumption that, if each individual has some independent ability to acquire valuable information, then any idea or behavior that is widely accepted is likely to be worth adopting.

It is possible to imagine many other such heuristics. For instance, Henrich and his colleague Francisco Gil-White have suggested using a variation of the conformity bias to improve on the success bias.[37] They point out that even figuring out who is successful can be difficult. For instance, in small-scale societies, which hunter brings in the most game varies widely from one day to the next.[38] In the midst of this statistical noise, how can we decide which hunter to imitate? We can turn to others. If many people look up to a given individual—if that individual has prestige—then imitating them might be worthwhile. For Henrich and Gil-White, such a *prestige bias* is highly adaptive.

Boyd, Richerson, Henrich, and others have built sophisticated models showing how reliance on rough heuristics allows individuals to make the best of their surrounding culture. Another advantage of these heuristics is that they are cognitively cheap, with no need for complex cost-benefit calculations: figure out what most people believe and adopt the same beliefs, or figure out who does something best and imitate everything they do.[39]

But what happens when the majority is wrong, or when the most successful or prestigious individual was just lucky? If these rough heuristics provide a good bang for the buck—decent results at a cheap cost—they also lead to systematic mistakes.

Boyd, Richerson, and Henrich are ready to bite the bullet. The self-sacrifice of the Japanese kamikaze is accounted for through a type of conformity bias, which allows cultural elements that are beneficial for the group, but detrimental to the individual, to spread.[40] The prestige bias would explain why people appear more likely to kill themselves after a celebrity has committed suicide.[41] Less dramatically, success bias predicts that people will buy underwear advertised by basketball star Michael Jordan, even though his athletic prowess is likely unrelated to his taste in undergarments.[42]

THE CASE FOR GULLIBILITY   13

Not only do gene-culture coevolution theorists bite the bullet, but they do so gleefully. They accept that "to get the benefits of social learning, humans have to be credulous, for the most part accepting the ways that they observe in their society as sensible and proper."[43] Indeed, the fact that reliance on rough heuristics predicts the spread of absurd beliefs and maladaptive behavior, as well as useful ones, is an "interesting evolutionary feature of these rules."[44] The novelty of this idea—maladaptive culture spreads because we are adapted for culture—makes it all the more attractive.

THE CASE AGAINST GULLIBILITY

Many theories in the social sciences can be roughly recast in the terms of this gene-culture coevolution framework. "The ideas of the ruling class are in every epoch the ruling ideas," as Marx and Engels suggested: success bias.[45] People blindly follow the majority: conformity bias. Charismatic leaders go from being worshipped by their faction to controlling the masses: prestige bias. An incredible array of intellectual traditions—centuries-old political philosophy, experimental psychology, biologically inspired modeling—converge on the notion that humans are, by and large, credulous, overly deferential toward authority, and excessively conformist.

Could this be all wrong?

Throughout this book, I will chip away at the support for the idea that the masses are gullible. Here's the argument in a nutshell.

Once we take strategic considerations into account, it becomes clear that gullibility can be too easily taken advantage of, and thus isn't adaptive. Far from being gullible, humans are endowed with dedicated cognitive mechanisms that allow them to carefully evaluate communicated information. Instead of blindly following prestigious individuals or the majority, we weigh many cues to decide what to believe, who knows best, who to trust, and what to feel.

The multiple mass persuasion attempts witnessed since the dawn of history—from demagogues to advertisers—are no proof of human gullibility. On the contrary, the recurrent failures of these attempts attest to the difficulties of influencing people en masse.

Finally, the cultural success of some misconceptions, from wild rumors to supernatural beliefs, isn't well explained by a tendency to be credulous. By and large, misconceptions do not spread because they are pushed by prestigious or charismatic individuals—the supply side. Instead, they owe their success to demand, as people look for beliefs that fit with their preexisting views and serve some of their goals. Reassuringly, most popular misconceptions remain largely cut off from the rest of our minds and have few practical consequences, explaining why we can be relatively lax when accepting them.

2

VIGILANCE IN COMMUNICATION

THE BEST ARGUMENT IN FAVOR of credulity is that it enables us to acquire the knowledge of our peers and forebears. Being disposed to copy what others do or think, and relying on simple heuristics to figure out who to copy—what most people, or prestigious leaders, do and think—would give us easy access to a wealth of accumulated wisdom.

This argument, however, fails to take into account the strategic element present in all interactions. It assumes that the individuals being copied are doing their best to engage in adaptive behaviors and to form accurate beliefs. It does not consider that these individuals might wish to influence those who copy them. But why wouldn't they? Being able to influence others is a great power. And, from an evolutionary point of view, with great power comes great opportunities.

To understand what happens when individuals evolve to influence others as well as to be influenced by them, the most suitable framework is that of the evolution of communication. The counterintuitive predictions that follow from this theory are best illustrated with some puzzling animal behaviors, which I now describe, their explanations unfolding throughout the chapter.

ANIMALS ACTING WEIRD

In the woodlands of eastern Australia, one sometimes stumbles upon strange construction: tiny house-like structures made of grass, decorated with berries, eggshells, bits of metal, and sundry colorful objects. These constructions—called *bowers*—are not made by the local human population but by spotted bowerbirds. Do the birds use these painstakingly constructed structures for protection against the weather or against predators? No. They build more typical nests in trees for this purpose. Why do bowerbirds bother building bowers?

With a sleek shape, long horns, elegant black streaks on their flanks, and a bright white rump, Thomson's gazelles are gorgeous animals—gorgeous, but maybe a bit daft. Packs of wild dogs roam the savanna, ready to chase and devour them, yet when a gazelle spots a pack of dogs, it often fails to flee at full speed. Instead, it stots, jumping on the same spot while keeping its legs straight. It stots high, sometimes as high as six feet.[1] It stots in the absence of any obstacles. It stots even though stotting slows it down. Why does the stupid gazelle not stop stotting?

Like the spotted bowerbird, the Arabian babbler is a brownish bird, just short of one foot long. As the name suggests, though, babblers do not build bowers; they babble. Besides its vocal displays, a striking feature of the Arabian babbler is its cooperativeness: groups of a dozen babblers look after their offspring together, clean each other, and act as sentinels. When they see a predator approaching, these sentinels engage in a behavior that seems much more sensible than that of our gazelles: they give alarm calls. When the predator is still quite distant, the sentinels emit barks (two relatively low calls) or trills (a higher, longer vibrating call). When the predator draws closer, the sentinels start emitting tzwicks (three short, higher-pitched calls). These calls allow the other group members to hide from some predators and to mob others. So far so good. Some babblers, however, have solitary lives: they do not live, let alone cooperate, with other babblers. Yet when these so-called floaters spot a predator, they give the same calls as the sentinels.[2] Why would lonely babblers give such wanton warning wails?

In humans, as in other mammals, pregnancy causes many changes in a mother's body. Some are obvious—the enlarged belly—but others are subtler, such as the change in the way insulin is produced. Insulin is the hormone that tells the body to turn blood sugar into fat. After a sugar-rich meal, blood sugar levels increase, insulin is secreted, and the sugar is stored away in the form of fat. Toward the end of pregnancy, however, mothers start generating increasingly large quantities of insulin after their meals. This might seem weird: the growing fetus asks for a huge amount of energy, which it draws from the sugar in the mother's blood. Even weirder: in spite of this huge spike in insulin, blood sugar levels remain elevated for longer than usual.[3] Why would a mother's body labor to limit the little one's resources? And why would it fail?

For animals with such tiny brains, bees are very sophisticated foragers. They scout for nectar-rich flowers and keep track of their location. When they return to the hive, the bees use their famous waggle dance to tell their hivemates where to find food. To forage effectively, bees are apt to use both personal experience (where they have found good flower spots in the past) and social information (the dances of other bees). In order to test how much weight bees put on personal versus social information, entomologist Margaret Wray and her colleagues performed a series of ingenious experiments. They put a feeder (an artificial source of sugar) in the middle of a lake. Some bees flew over the lake, found the feeder, and returned to the hive with the good news. Now, flowers do not grow in the middle of lakes. The bees back at the hive, upon seeing a dance pointing at the lake, would have been quite justified in thinking the returning bees mistaken. But they did not. The bees dutifully left the hive—indeed, they left the hive at the same rate as when the feeder had been placed in a much more plausible location.[4] Why would these intelligent insects ignore their individual intuitions and follow implausible instructions?

CONFLICTS AND THE EVOLUTION OF COMMUNICATION

The key to understanding these strange behaviors is also the key to understanding how we evaluate what we are told: the theory of the evolution of communication.

For genuine communication to exist, there must be dedicated adaptations on the side of the entity sending the signals, and on the side of the entity receiving the signals.[5] For example, vervet monkeys have a sophisticated system of alarm calls, allowing them to warn each other of the presence of eagles, snakes, leopards, and other predators. Vervets must be equipped with mechanisms that trigger the correct call when each of these predators has been spotted, as well as with mechanisms that trigger the appropriate reaction to each of these calls—climbing trees is not very helpful when an eagle draws near.[6] These alarm calls clearly fit the bill for an evolved communication system.

If one side is endowed with specific adaptations, either to emit or to receive some information, without a counterpart on the other side, there is no genuine communication. Instead of communication, there can be cues, which only require adaptations on the receiving side. For instance, adult mammals can differentiate babies from adult members of their species. But they do not need communication to do so; they can rely on cues—most obviously, size. Babies did not evolve to be small so that they could be recognized as babies. Small size is a cue to babyhood, not a signal.

Now, the theory of evolution by natural selection dictates that if communication mechanisms have evolved, they must have done so because they increased the fitness of both the entities that send signals and the entities that receive signals. Fitness, in evolutionary theory, is the reproductive success of an entity—which includes not only its own reproduction but also the reproduction of copies of itself. So individuals can increase their fitness by having more descendants, but also by helping their kin, who are more likely to share any new gene variant the individual possesses, have more descendants—what biologists call *inclusive fitness*.

In some cases, the evolution of communication is straightforward enough. The cells of an individual share the same fitness: the cells of your liver and your brain both increase their fitness when you reproduce. Their interests are perfectly aligned with each other's. As a result, there is no reason for a cell to mistrust what another cell from the same body might communicate, no obstacle to the evolution of communication between them. Indeed, our cells keep listening to each other even when some of them turn bad: cancerous cells emit signals that tell the body to grow more blood vessels, and the body obeys.[7]

Entities can also share the same fitness without being part of a single body. The fitness of worker bees, for instance, is entirely tied with the reproductive success of the queen. The workers cannot reproduce on their own, and their only chance of passing on their genes is through the queen's offspring. As a result, worker bees have no incentive to deceive one another, and this is why a bee can trust what another bee signals without having to double-check—even if the bee suggests there are flowers in the middle of a lake.

Still, a lot of communication happens between individuals that don't share the same fitness. In these potentially conflicting interactions, many signals might improve the fitness of senders, while doing nothing for the fitness of receivers, or even decreasing their fitness. For instance, a vervet monkey might give out an alarm call not because there is a predator in sight but because it has spotted a tree laden with ripe fruit and wants to distract the other monkeys while it gorges on the treat. We might refer to such signals as dishonest or unreliable signals, meaning that they are harmful to the receivers.

Unreliable signals, if they proliferate, threaten the stability of communication. If receivers stop benefiting from communication, they evolve to stop paying attention to the signals. Not paying attention to something is easily done. If a given structure is no longer advantageous, it disappears—as did moles' eyes and dolphins' fingers. The same would apply to, say, the part of our ears and brains dedicated to processing auditory messages, if these messages were, on balance, harmful to us.

Likewise, if receivers managed to take advantage of senders' signals to the point that the senders stopped benefiting from communication, the senders would gradually evolve to stop emitting the signals.[8] Communication between individuals that do not share the same incentives—the same fitness—is intrinsically fragile. And the individuals don't have to be archenemies for the situation to degenerate.

SURPRISING COMMUNICATION FAILURES

We tend to think of pregnancy as a symbiotic relationship between the mother and her offspring. In fact this relation is, to some extent, conflicting from the start. To maximize their own fitness, mothers should not dedicate all of their resources to the fetus they are currently carrying. Instead, some resources should be devoted to past and future children (and thus to the mother herself). By contrast, the fetus should evolve to be biased toward itself, compared to its siblings. As a result of this asymmetry in the selective pressures bearing on mothers and their fetuses, the fetuses should evolve to ask more resources of the mothers than the mothers would optimally allocate to any one offspring.

David Haig, a particularly inventive evolutionary biologist, suggested that this difference of selective pressures between mothers and fetuses explains, among many other phenomena, the strangeness of pregnant mothers' insulin physiology.[9] Through the placenta, the fetus produces and releases hormones into the mother's blood. One of these hormones, human placental lactogen (hPL), increases insulin resistance. The more resistant the mother is to insulin, the longer her blood sugar remains elevated, and the more resources the fetus can grab. In response, the mother increases her production of insulin. At the end mother and fetus reach a kind of equilibrium in which blood sugar remains elevated slightly longer than usual, but much less longer than would be the case if the mother didn't release increased doses of insulin. The efforts made by the fetus to manipulate its mother's blood sugar levels are staggering: the placenta secretes one to three grams of hPL per day.[10] For a tiny organism that should be busy growing, this is a significant expenditure of resources. By comparison, placental hormones that are not subject to this tug-of-war can affect the mother with doses a thousand times smaller.

If evolutionary logic makes sense of some bizarre phenomena, like a mother and her fetus using hormones to battle over resources, it also opens up new dilemmas. Take alarm calls. Until the 1960s, their function was pretty much taken for granted: an individual gives these calls to warn its fellow group members. The belief was that even if giving alarm calls means spending time on the lookout rather than, say, feeding, as well as being more vulnerable to predation, it is worthwhile, since it increases the odds that the group will survive. In his classic *Adaptation and Natural Selection*, published in 1966, biologist George Williams forcefully argued against this logic. Imagine that one individual in the group evolves to not give alarm calls, or to give them less often. This individual is better off than everyone else: it still benefits from the others' warnings but pays a lesser cost, or no cost at all, in return. This trait will be selected and spread in the population, until no one sounds the alarm anymore. So why do alarm calls persist in many species? In some cases, the answer is to be found in kin selection. For example, yellow-bellied marmots give alarm calls, but not all of them do so equally. Most of the alarm calls are given by mothers that have just had new pups. The pups, not being as good as older individuals at spotting predators, are likely to benefit substantially from the calls. Warning their pups is a good investment for mothers, who don't bother warning other group members.[11]

A similar phenomenon might explain some alarm calls in Arabian babblers: they live in groups of highly related individuals, and the calls might boost the fitness of the caller by helping its offspring, or the offspring of its siblings, survive.[12] But that does not explain why floaters—solitary babblers—also give out alarm calls, even though they have no one to warn.

SURPRISING COMMUNICATION SUCCESSES

The logic of the evolution of communication explains why it can be so hard for individuals who have much in common—a mother and her fetus—to communicate efficiently. It also explains how communication emerges between individuals who seem to be locked in purely adversarial relationships. Even though the existence and the extent of common incentives matter, what matters even more is the possibility—or lack thereof—of keeping signals honest and thus mostly beneficial for those who receive them.

What common incentives do predator and prey have? Neither wants to waste resources. If a prey is nearly certain to escape its predator, they are both better off if the predator doesn't attack at all, and they can both save some energy. But prey can't simply send predators a signal meaning "You can't catch me!" All prey would have an incentive to send this signal, even if they were too young, old, tired, hurt, or unprepared to escape the predator. Predators, then, would have no reason to believe the signal. For such a signal to function and to last, it should be disproportionally likely to come from prey fit enough to escape. Otherwise, it is not evolutionarily stable, and so it will be selected out and eventually disappear (or never appear in the first place).

This might be what the alarm call of the Arabian babbler achieves. By giving an alarm call, a babbler tells the predator that it has been spotted. Once it has been spotted, the predator's chances of launching a successful attack are low, as the babbler can now seek cover. Many species, from lizards to kangaroo rats, warn predators in this way.[13] What keeps the signal honest, guaranteeing its evolutionary stability? Why don't babblers emit these calls at frequent intervals, just in case there happens to be a predator around? One reason is that the calls don't always deter predators; they simply lower the odds of an attack. If the prey has already been spotted by a predator, giving the call is worthwhile. But if the prey hasn't been spotted, then it just made its position known to any predators nearby and, since it doesn't know where these predators might be, its chances of escape are low. As a result, prey have an incentive to give the calls only when they have actually spotted the predator, making the calls credible. Predator-deterrent signals have some intrinsic credibility, but they can be made even more convincing by the prey orientating toward the predator in a very visible manner—something the prey could not do without having already spotted the predator, making the signal even more credible.[14] For instance, Thomson's gazelles turn their rump toward a predator when they have spotted one. The rump has a white patch, making it easier for predators to get the signal.[15] In this way, the gazelle can signal to its predator that it has been spotted while still facing the other way in case the predator decides that rump is too appetizing to pass up on.

Thomson's gazelles not only show their rump to predators. They also stot. Far from being useless, these high jumps also act as a predator-deterrent signal. The gazelles are telling predators that they are so fit that they would be sure to outrun them, so why bother? Stotting is a reliable signal because only a fit gazelle can stot enough, and high enough, to dissuade predators.

Stotting provides a good illustration of the kind of evidence used to test an evolutionary hypothesis. How can we know that the main function of stotting in Thomson's gazelles is to deter predator pursuit? First, we can rule out some alternative hypotheses. Stotting does not increase the gazelles' speed; indeed, they stop stotting when the predators get too close.[16] Stotting isn't used to avoid obstacles, since gazelles usually stot even though there's nothing in the way. Stotting doesn't simply signal to the predators that they have been spotted, since the gazelles rarely stot when they spot a cheetah. Cheetahs are ambush predators and thus don't care about the gazelle's ability to sustain a long race.

Having ruled out these alternatives, what positive evidence is there in favor of the hypothesis that stotting has the function

VIGILANCE IN COMMUNICATION 25

of deterring predator pursuit? First, the gazelles stot in response to the right predators: wild dogs, which are coursers. This makes sense if they are advertising their ability to run fast for a long time. Second, the gazelles stot more when they are fitter (in the wet season) than when they are less fit (in the dry season). Third, stotting works: wild dogs are less likely to chase gazelles that stot more, and once the chase has begun, they are more likely to switch toward gazelles that stot less.

HOW TO SEND COSTLY SIGNALS FOR FREE

Natural selection has stumbled on impressively creative ways of keeping communication honest, even in very adversarial relationships, by making it essentially impossible to send unreliable signals. Babblers can't coordinate their calls with those of unseen predators. Only a fit gazelle can stot convincingly. Yet, human beings do not seem to have any comparable way of demonstrating the reliability of the messages they send. With a few anecdotal exceptions—such as saying "I'm not mute," which reliably communicates that one isn't mute—there are no intrinsic constraints on sending unreliable signals via verbal communication. Unlike an unfit gazelle that just can't stot well enough, a hack is perfectly able to give you useless advice.

A commonly invoked solution to keep human communication stable is costly signaling: paying a cost to send a signal would be a guarantee of its reliability. Costly signaling supposedly explains many bizarre human behaviors. Buying luxury brands would be a costly signal of wealth and status.[17] Constraining religious rituals—from frequent public prayer to fasting—would be a costly signal of one's commitment to a religious group.[18] Performing dangerous activities—from turtle hunting among the Meriam hunter-gatherers to reckless driving among U.S. teens—would be costly signals of one's strength and competence.[19]

Costly signaling is often invoked but often misunderstood. Intuitively, what matters for a costly signal to work is the cost paid by those who send the reliable signal: it would be because someone pays more than a thousand dollars to buy the latest iPhone that owning this phone is a credible signal of wealth. In fact, what matters is that, compared with reliable signalers, unreliable signalers incur a greater cost when sending the signal. In other words, what matters isn't the cost of buying the new iPhone per se but the fact that spending so much money on a phone is costlier for a poor person, who might have to skimp on necessities to afford it, than for a rich person, for whom a thousand dollars might make very little difference.[20]

Given that what matters is a difference—between the costs of sending a reliable and an unreliable signal—the absolute level of the cost doesn't matter. As a result, costly signaling can, counterintuitively, make a signal reliable even if no cost is paid. As long as unreliable signalers would pay a higher cost if they sent signals, reliable signalers can send signals for free. The bowerbirds' bowers illustrate this logic.

It is now well accepted that male bowerbirds build their bowers to attract females. Indeed, better-decorated bowers succeed in providing their builder more mating opportunities.[21] Why would females be attracted by fancy bowers? After all, these bowers are of absolutely no practical use. Instead, Amotz Zahavi, a biologist who did much to develop the theory of costly signaling, suggested that bowerbirds indicate their value as mates by showing they can pay the cost of building fancy bowers—which might require taking risks, or going hungry while they forage for fancy frills instead of tasty treats.[22] In fact, building bowers doesn't appear to be particularly costly: males are not more likely to die during the bower-building season, in spite of their construction efforts.[23] So what makes the bower a reliable signal?

The mechanism was discovered, somewhat inadvertently, by ornithologist Joah Madden when he attempted to trick female bowerbirds by putting extra berries on some of the bowers.[24] Typically, the females prefer to mate with the males that build the most berry-rich bowers. But the berries added by Madden had no such effect. It wasn't that he had bad taste in berries, from a bowerbird perspective. Instead, rival bowerbirds were sabotaging the bowers that had received Madden's gift of berries. The other bowerbirds took these extra berries to mean that the bowers' owners were pretending to be of higher status than they actually were, and they vandalized the bowers to put the owners back in their place.

What keeps the system stable isn't the intrinsic cost of building a fancy bower (which is low in any case). Instead, it is the vigilance of the males, who keep tabs on each other's bowers and inflict a cost on those who build exaggerated bowers. As a result, as long as no male tries to build a better bower than they can afford to defend, the bowers send reliable signals of male quality without any significant cost being paid. This is costly signaling for free (or nearly for free, as there are indirect costs in monitoring other males' bowers).

As we will see, this logic proves critical to understanding the mechanisms that allow human communication to remain stable. No intrinsic cost is involved in speaking: unlike buying the latest iPhone, making a promise is not intrinsically costly. Human verbal communication is the quintessential "cheap talk" and thus seems very far from qualifying as a costly signal. This is wrong. What matters isn't the cost borne by those who would keep their promises but the cost borne by those who do not keep them. As long as there is a mechanism to exert a sufficient cost on those who send unreliable messages—if only by trusting them less in the future—we're dealing with costly signaling, and communication can be kept stable. Undoubtedly, the fact that humans have developed ways of sending reliable signals without having to pay a cost every time they do so has greatly contributed to their success.

THE NEED FOR VIGILANCE

Communication is a tricky business. We find successes and failures of communication in the most surprising places: prey can convince predators to give up chasing them, a fetus can't persuade its mother to give it more resources. The logic of evolution is crucial to understanding these successes and failures. It tells us when individuals have common incentives: cells in a body, bees in a beehive. But as the conflicts occurring during pregnancy illustrate, having some common incentives isn't enough. Unless the reproductive fates of two entities are perfectly intertwined, incentives to send unreliable signals are pretty much guaranteed to exist. This is when natural selection gets creative, having stumbled on a variety of ways to keep signals reliable. Some of these solutions—such as the gazelles' stots—are fascinating but hardly applicable to human communication. Instead, I will argue that human communication is kept (mostly) reliable by a whole suite of cognitive processes—mechanisms of open vigilance—that minimize our exposure to unreliable signals and, by keeping track of who said what, inflict costs on unreliable senders.

How these mechanisms work, how they help us decide what to believe and who to trust, is the topic of the next five chapters.

What should be clear in any case is that we cannot afford to be gullible. If we were, nothing would stop people from abusing their influence, to the point where we would be better off not paying any attention at all to what others say, leading to the prompt collapse of human communication and cooperation.

3

EVOLVING OPEN-MINDEDNESS

FOR HUMANS, THE ABILITY to communicate is of enormous significance. Without communication, we would have a hard time figuring out what we can safely eat, how to avoid danger, who to trust, and so forth. Although effective communication is arguably more important than ever, it was also critical for our ancestors, who needed to communicate with each other in order to hunt and gather, to raise their children, to form alliances, and to pass on technical knowledge.[1] Our complex vocal and auditory apparatuses, which clearly serve sophisticated verbal communication, are at least as old as anatomically modern humans—three hundred thousand years. That our cousins the Neanderthals, from whom our ancestors split more than six hundred thousand years ago, appear to have had the same anatomical equipment, suggests complex verbal communication is considerably older.[2]

If, from a very early stage in their (pre)history, humans stood to gain enormous benefits by communicating with each other, they have also been at risk from the abuse of communication. More than any other primate species, we are in danger of being misled and manipulated by communication. The existence of an evolutionarily relevant problem creates selection pressures that favor the development of cognitive mechanisms dedicated to solving this problem. The same is true of communication, with its promises and its perils.

Indeed, the stakes are so high that it would be puzzling if we hadn't evolved specialized cognitive mechanisms to deal not only with the potential but also with the danger of communication. In an article from 2010, cognitive scientist Dan Sperber and some colleagues (including yours truly) called these mechanisms *epistemic vigilance*, but I will call them here *open vigilance*, to stress that these mechanisms are at least as much about being open to communicated information as being vigilant toward it.[3] However, even if we agree such mechanisms must exist, there are different manners in which they could function.

One way of thinking about the evolution of communication, and thus our open vigilance mechanisms, is to use the arms race analogy. An arms race is a competition between two entities in which each, in response to the other's move, progressively ups the ante. The analogy emerged during the Cold War, as Russia and the United States built more nuclear weapons as a reaction to the other power building more nuclear weapons, which was a reaction to the other power building more nuclear weapons, and so on.

In the case of communication, an arms race could take place between senders, which evolve increasingly sophisticated means of manipulating receivers, and receivers, which evolve increasingly sophisticated means of rejecting unreliable messages. This is what we get, for instance, with computer viruses and security software. In the human case, this model leads to an association between lack of mental acuity and gullibility. Many commentators throughout history have suggested that some humans—from women to slaves—have stringent intellectual limitations, limitations that would make these populations gullible (in my terms, by precluding the use of more sophisticated mechanisms of open vigilance). Even assuming we are all endowed with the same cognitive equipment, we are not always able to rely on it. Thus the arms race model predicts that when receivers, because they are exhausted or distracted, cannot use properly their most refined cognitive mechanisms, they would be defenseless against the senders' more advanced cognitive devices, in the same way that a security software system that has not been updated leaves a computer vulnerable to attacks.

BRAINWASHERS AND HIDDEN PERSUADERS

For America in the 1950s, fear of manipulation was in the zeitgeist. With Joseph Stalin still at the helm of the Soviet Union, the perceived communist threat was at its height, and the United States had reached peak McCarthyism. The "Reds" were thought to have infiltrated everything: the government, academia, defense programs. Even more insidiously, they were supposed to have wormed their way into the minds of that most dedicated, most patriotic American: the soldier. During the Korean War, thousands of U.S. soldiers were captured by the Koreans and the Chinese. Those who managed to escape brought back tales of horrible mistreatment and torture, from sleep deprivation to waterboarding. When the war ended, and the prisoners of war (POWs) were repatriated, these mistreatments acquired an even darker meaning. Not simply an example of the wanton cruelty of the enemy, they were seen as an attempt to brainwash U.S. soldiers into accepting communist doctrine. Twenty-three American POWs chose to follow their captors to China instead of going back to their homeland, which was surely, as the *New York Times* stated at the time, "living proof that Communist brainwashing does work on some persons."[4]

Brainwashing was supposed to function by shattering people's ability for higher reflection, as it involved "conditioning," "debilitation," and "dissociation-hypnosis-suggestibility."[5] For U.S. rear admiral Daniel Gallery, it made of men "borderline case[s] between a human being and a rat struggling to stay alive."[6] The techniques used by the Koreans and Chinese were thought to be derived from those developed earlier by the Russians, turning the POWs into "prisoners of Pavlov,"[7] referring to the psychologist famous for making dogs salivate at the sound of a bell. Ironically, Americans in their "war on terror" would go on to use many of the same techniques—waterboarding being a case in point—when attempting to extract information from suspected terrorists.

In 1950s America, the idea that people are more easily influenced when they cannot think also showed up in a very different context. The targets weren't POWs suffering the hell of Korean prison camps, but moviegoers comfortably watching the latest Hollywood blockbusters. In the midst of the movie, messages such as "drink Coke" were presented so quickly that they could not be consciously perceived.[8] These messages would soon be called *subliminal*, meaning "below the threshold," here the threshold of awareness. Subliminal messages created a scare that would last for decades. As late as 2000, a scandal erupted when a Republican-funded advertisement that attacked policy proposals made by Al Gore—the Democratic candidate for the U.S. presidency—was found to have presented the word *rats* subliminally to its viewers.[9] The power of subliminal messages was also harnessed for nobler causes. Companies started producing therapeutic tapes—to enhance self-esteem, say—that people could listen to in their sleep. Because people don't tend to exercise much conscious control when they are asleep, the recordings were aimed directly at their subconscious and were thus believed to be particularly effective.

The scares surrounding brainwashing and subliminal influence rely on a pervasive association between inferior cognitive ability and gullibility: the less we think, the worse we think, and the more we will be influenced by harmful messages. This association between lack of intellectual sophistication and gullibility is historically pervasive. Already when Heraclitus, in 500 BCE, talked about "the people" who "let themselves be led by speechmakers, in crowds, without considering how many fools and thieves they are among [them]," he was talking about the masses, the common people, not the aristocrats.

Twenty-five centuries later, the same trope pervaded the discourse of crowd psychologists. These European scholars, working in the latter half of the nineteenth century, grappled with the growing impact of crowds in politics, from revolutionary mobs to striking miners. These scholars developed a view of crowds as both violent and gullible that would prove immensely popular, inspiring Benito Mussolini and Adolf Hitler and being, to this day, common among those who have to deal with crowds, such as members of law enforcement agencies.[10] The best known of these crowd psychologists, Gustave Le Bon, suggested that crowds shared the "absence ... of critical thought ... observed in beings belonging to inferior forms of evolution, such as women, savages, and children."[11] In a beautiful illustration of motivated reasoning, Le Bon's colleague Gabriel Tarde claimed that because of its "docility, its credulity ... the crowd is feminine" even when, as he admitted, "it is composed, as is usually the case, of males."[12] Another of the crowd psychologists, Hippolyte Taine, added that in crowds, people were reduced to the state of nature, like "servile monkeys each imitating the other."[13] At about the same time, on the other side of the Atlantic, Mark

Twain was depicting Jim as a "happy, gullible, rather childlike slave."[14]

In the twenty-first century, we still find echoes of these unsavory associations. Writers for the *Washington Post* and *Foreign Policy* claim Donald Trump was elected thanks to the "gullibility" of "ignorant" voters.[15] A common view of Brexit—the vote for Britain to leave the European Union—is to see the Brexiters as "uneducated plebs" while those who voted remain are "sophisticated, cultured and cosmopolitan."[16]

In contemporary academic literature, the link between unsophistication and credulity mostly takes two forms. The first is in children, whose lack of cognitive maturity is often associated with gullibility. A recent psychology textbook asserts that as children master more complex cognitive skills, they become "less gullible."[17] Another states, more sweepingly, that "children, it seems, are an advertiser's dream: gullible, vulnerable, and an easy sell."[18]

The second way in which lack of cognitive sophistication and credulity are linked is through a popular division of thought processes into two main types, so-called System 1 and System 2. According to this view—long established in psychology and recently popularized by psychologist Daniel Kahneman's *Thinking, Fast and Slow*—some cognitive processes are fast, effortless, largely unconscious, and they belong to System 1. Reading a simple text, forming a first impression of someone, navigating well-known streets all belong to System 1. The intuitions that form System 1 are, on the whole, effective, yet they are also susceptible to systematic biases. For instance, we seem to make judgments of people's competence or trustworthiness on the basis of facial traits. These judgments may have some limited reliability, but they should be easily superseded by stronger cues—such as how the person actually behaves.[19] This is when

System 2 is supposed to kick in. Relying on slow, effortful, reflective processes, System 2 takes over when System 1 fails, correcting our mistaken intuitions with its more objective processes and more rational rules. This is the common dual-process narrative.[20]

Maybe the best-known task exemplifying the function of the two systems is the Bat and Ball:

> A bat and a ball cost $1.10 in total. The bat costs $1.00 more than the ball. How much does the ball cost?[21]

If you haven't encountered it before, give it a shot before reading further.

This problem has fascinated psychologists because, in spite of its seeming simplicity, most people provide the wrong answer—namely, ten cents. The ten cents answer is the perfect example of a System 1 answer: for the majority of people, it is the first thing that pops into their heads after they have read the problem. Yet ten cents cannot be correct, for then the bat would cost $1.10, and the two together $1.20. Most people have to rely on their System 2 to correct this intuitive mistake, and to reach the correct answer of five cents.[22]

If System 1 consists of rough-and-ready mechanisms, while System 2 consists of slow, deliberate reflection, we might expect System 1 to be associated with credulity, and System 2 with critical thinking. Psychologist Daniel Gilbert and his colleagues performed an ingenious series of experiments to tease out the role played by the two mental systems in the evaluation of communicated information.[23] In these experiments, participants were presented with a series of statements, and right after each statement was presented, they were told whether it was true or false. For instance, in one experiment, the statements were about words in Hopi (a Native American language), so participants might be told "A ghoren is a jug" and, a second later, they were told "true." After all the statements had been presented, participants were asked which had been true and which had been false. To test for the role played by the two systems, Gilbert and his colleagues intermittently interrupted System 2 processing. System 2, being slow and effortful, is easily disrupted. In this case, participants simply had to press a button when they heard a tone, which tended to ring when the crucial information—whether a given statement was true or false—was being delivered.

When it came time to recall which statements were true and which were false, people whose System 2 had been disrupted were more likely to believe the statements to be true—irrespective of whether they had in fact been signaled as true or false. The System 2 disruption had thus caused many participants to accept false statements as true. These experiments led Gilbert and his colleagues to conclude that our initial inclination is to accept what we are told, and that the slightest disruption to System 2 stops us from reconsidering this initial acceptance. As Gilbert and his colleagues put it in the title of their second article on the topic: "You Can't Not Believe Everything You Read."[24] Kahneman summarized these findings as follows: "When System 2 is otherwise engaged, we will believe almost anything. System 1 is gullible and biased to believe, System 2 is in charge of doubting and unbelieving, but System 2 is sometimes busy, and often lazy."[25]

These results are in line with the associations observed between a more "analytic" thinking style—that is, being more inclined to rely on System 2 than System 1—and the rejection of empirically dubious beliefs. In a widely publicized article, psychologists Will Gervais and Ara Norenzayan found that people with a more analytic frame of mind—people who are better at solving problems like the Bat and Ball, for instance—are more likely to be atheists.[26] Other studies suggest that more analytically inclined participants are less likely to accept a variety of paranormal beliefs, from witchcraft to precognition.[27]

I HOPE YOU DID NOT BELIEVE EVERYTHING YOU JUST READ

The association between lack of cognitive sophistication and gullibility, predicted by the arms race view of the evolution of vigilance, has been prevalent throughout history, from Greek philosophers to contemporary psychologists. Yet, as appealing as they might be, I believe that the arms race analogy, along with the association between lack of sophistication and gullibility, are completely mistaken, with critical consequences for who is more likely to accept wrong beliefs, and why.

For a start, the arms race analogy doesn't fit the broad pattern of the evolution of human communication. Arms races are characterized by the preservation of the status quo through parallel escalation. Russia and the United States acquired increasingly large nuclear arsenals, but neither nation gained the upper hand. Computer viruses haven't been wiped out by security software, but the viruses haven't taken over all computers either. Likewise, in the fight for resources between the mother and her fetus described in the previous chapter, the increasingly large deployment of hormonal signals on both sides has practically no net effect.

Human communication is, fortunately, very different from these examples. Here, the status quo might be the amount of information our prehuman ancestors or, as an approximation, our closest living relatives, exchange. Clearly, we have ventured very, very far from this status quo. We send and consume orders of magnitude more information than any other primate, and, crucially, we are vastly more influenced by the information we receive. The bandwidth of our communication has dramatically expanded. We discuss events that are distant in time and space; we express our deepest feelings; we even debate abstract entities and tell stories about imaginary beings.

For the evolution of human communication, a better analogy than the arms race is the evolution of omnivorous diets. Some animals have extraordinarily specific diets. Koalas eat only eucalyptus leaves. Vampire bats drink only the blood of live mammals. Pandas eat only bamboo. These animals reject everything that isn't their food of choice. As an extreme example, koalas will not eat a eucalyptus leaf if it isn't properly presented—if it is on a flat surface, for example, rather than attached to the branch of a eucalyptus tree.[28] These animals have evolved extremely specific food choices. However, this strategy can backfire if they find themselves in a new environment. Vampire bats drink only the blood of live mammals, so they don't have to worry about whether their food is fresh. Because the problem of learning to avoid toxic food is not one they face in their natural environment, they have no mechanism for learning food aversion, and keep drinking food that they should associate with sickness.[29]

By contrast with these specialists, omnivorous animals are both more open and more vigilant. They are more open in that they search for, detect, and ingest a much wider variety of foods. Rats or humans need more than thirty different nutrients, including "nine amino acids, a few fatty acids, at least ten vitamins, and at least thirteen minerals,"[30] and none of their food sources can provide all of those at once. Omnivores have to be much more open in the range of foods they are willing to sample. Indeed, rats or humans will try just about anything that looks edible. They are endowed with a suite of mechanisms that detects the various nutrients they need in what they ingest, and adjust their diet according to these needs—craving salty foods when low in sodium, and so forth.[31]

This openness makes omnivores fantastically adaptable. Humans have been able to survive on diets made up almost exclusively of milk and potatoes (early eighteenth-century Irish peasants) or meat and fish (the Inuit until recently). However, their openness also makes omnivores vulnerable. Meat can go bad and contain dangerous bacteria. To avoid being eaten, most plants are either toxic or hard to digest. As a result, omnivores are also much more vigilant toward their food than specialists. Using a variety of strategies, they learn how to avoid foods that are likely to have undesirable side effects. The most basic of these strategies is to keep track of which foods made them sick and avoid these foods in the future—something that, as omnivores, we take for granted, but that some animals, such as vampire bats, are unable to do. Keeping track of which food is safe to eat requires some dedicated circuitry, not general learning mechanisms. The sick animal must learn to avoid the food it ate a few hours ago, and not all the other stimuli—what it saw, felt, smelled in between eating and getting sick.[32] Omnivores, from rats to humans, but also caterpillars, prefer food they have eaten when they were young.[33] Rats and humans also pay close attention to what other members of their species eat and whether or not it makes them sick, learning by observation which foods are safe.[34]

In terms of communication, the difference between humans and other primates is similar to the difference between specialists and omnivores. Nonhuman primates mostly rely on specific signals. Vervet monkeys have a dedicated alarm call for aerial predators;[35] chimpanzees smile in a way that signals submission;[36] dominant baboons grunt to show their pacific intentions before approaching lower-ranking individuals.[37] Humans, as noted earlier, are communication omnivores: they can communicate about nearly anything they can conceive of. Humans are thus vastly more open than other primates. Take something as basic as pointing. Human babies understand pointing shortly after they reach their first year.[38] But adult chimpanzees, even in situations in which pointing seems obvious to us, do not get it. Repeated experiments have put chimpanzees in front of two opaque containers, one containing food, but they don't know which. When an experimenter points to one of the containers, the chimpanzees are not more likely to pick this container than the other one.[39] It is not for lack of intelligence: if you try to grab one of the containers, the chimpanzees rightly infer that it must be the container with the food.[40] Communication is just much less natural for chimpanzees than it is for us.

If we are vastly more open to different forms and contents of communication than other primates, we should also be more vigilant. I will explore in the next four chapters how we exert this vigilance. Here I want to focus on the overall organization of our mechanisms of open vigilance. This organization is critical for understanding what happens when some of these mechanisms are impaired: Do such impairments make us more or less likely to accept misleading information?

According to the arms race theory, we have evolved from a situation of extreme openness, of general gullibility, toward a state of increasingly sophisticated vigilance made possible by our more recently developed cognitive machinery. If this machinery were removed, the theory goes, we would revert to our previous state of gullibility and be more likely to accept any message, however stupid or harmful.

The analogy with the evolution of omnivorous diets suggests that the reverse is the case. We have evolved from a situation of extreme conservatism, a situation in which we let only a restricted set of signals affect us, toward a situation in which we are more vigilant but also more open to different forms and contents of communication. This organization, in which increased sophistication goes with increased openness, makes for much more robust overall functioning. In the arms race view, disruption of the more sophisticated mechanisms makes us credulous and vulnerable. By contrast, a model in which openness and vigilance evolve hand in hand is not so fragile. If more recent mechanisms are disrupted, we revert to older mechanisms, making us less vigilant—but also much less open. If our more recent and sophisticated cognitive machinery is disrupted, we revert to our conservative core, becoming more stubborn rather than more gullible.[41]

BRAINWASHING DOES NOT WASH

What about the evidence that supports the association between lack of sophistication and gullibility and, indirectly, the arms race view of the evolution of vigilance? What about brainwashing and subliminal influence, for a start? If disrupting our higher cognitive abilities, or bypassing them altogether, were an effective means of influence, then both brainwashing and subliminal stimuli should leave us helpless, gullibly accepting the virtues of communism and thirsting for Coca-Cola. In fact, both persuasion techniques are staggeringly ineffective.

The brainwashing scare started when twenty-three American POWs defected to China after the Korean War. This is already a rather pitiful success rate: twenty-three converts out of forty-four hundred captive soldiers, or half a percent. But in fact the number of genuine converts was likely zero. The soldiers who defected were afraid of what awaited them in the United States. To gain some favors in the camps, they had collaborated with their Chinese captors—or at least had not shown as much defiance toward the captors as their fellow prisoners. As a result, those POWs could expect to be court-martialed upon their return. Indeed, among the POWs who had returned to the United States, one had been sentenced to ten years in jail, while prosecutors sought the death penalty for another. Compared with that, being feted as a convert to the Chinese system did not seem so bad, even if it meant paying lip service to communist doctrine—a doctrine they likely barely grasped in any case.[42] More recently, methods derived from brainwashing, such as the "enhanced interrogation techniques" that rely on physical constraints, sleep deprivation, and other attempts at numbing the suspects' minds, have been used by U.S. forces in their "war on terror." Like brainwashing, these techniques have been shown to be much less effective than softer methods that make full use of the suspects' higher cognition—methods in which the interrogator builds trust and engages the subjects in discussion.[43]

Similarly, the fear of subliminal influence and unconscious mind control was nothing but an unfounded scare. The early experiments demonstrating the power of subliminal stimuli were simply made up: no one had displayed a subliminal "drink Coke" ad in a movie theater.[44] A wealth of subsequent (real) experiments have failed to show that subliminal stimuli exert any meaningful influence on our behavior.[45] Seeing the message "drink Coke" flashed on a screen does not make us more likely to drink Coca-Cola. Listening to self-esteem tapes in our sleep does not boost our self-esteem. If some experiments suggest that stimuli can influence us without our being aware of it, the influence is at best minute—for instance, making someone who is already thirsty drink a little bit more water.[46]

What about the experiments conducted by Gilbert and his colleagues? They did show that some statements (such as "A ghoren is a jug") are spontaneously accepted and need some effort to be rejected. But does that mean that System 1 accepts "everything we read," as Gilbert put it? Not at all. If participants have some background knowledge related to the statement, this background knowledge directs their initial reaction. For instance, people's initial reaction to statements such as "Soft soap is edible" is rejection.[47] The statements don't even have to be obviously false to be intuitively disbelieved. They simply have to have some relevance if they are false. It is not very helpful to know that, in Hopi, it is not true that "a ghoren is a jug." By contrast, if you learn that the statement "John is a liberal" is false, it tells you something useful about John. When exposed to statements such as "John is a liberal," people's intuitive reaction is to adopt a stance of doubt rather than acceptance.[48] Far from being "gullible and biased to believe,"[49] System 1 is, if anything, biased to reject any message incompatible with our background belief, but also ambiguous messages or messages coming from untrustworthy sources.[50] This includes many messages that happen to be true. For instance, if you, like most people, got stuck on the ten-cents answer to the Bat and Ball problem, and someone had told you that the correct answer was five cents, your initial reaction would have been to reject their statement. In this case, your System 2 would have had to do some work to make you accept a sound belief, which is much more typical than cases in which System 2 does extra work to make us reject an unfounded belief.

There is no experimental evidence suggesting a systematic association between being less analytically inclined—less likely to use one's System 2—and being more likely to accept empirically dubious beliefs. Instead, we observe a complex interaction between people's inclinations to use different cognitive mechanisms and the type of empirically dubious beliefs they accept. Beliefs that resonate with people's background views should be more successful among those who rely less on System 2, whether or not these beliefs are correct. But an overreliance on System 2 can also lead to the acceptance of questionable beliefs that stem from seemingly strong, but in fact flawed, arguments.

This is what we observe: the association between analytic thinking and the acceptance of empirically dubious beliefs is anything but straightforward. Analytic thinking is related to atheism but only in some countries.[51] In Japan, being more analytically inclined is correlated with a greater acceptance of paranormal beliefs.[52] Where brainwashing techniques failed to convert any POWs to the virtues of communism, the sophisticated arguments of Marx and Engels convinced a fair number of Western thinkers. Indeed, intellectuals are usually the first to accept new and apparently implausible ideas. Many of these ideas have been proven right (from plate tectonics to quantum physics), but a large number have been misguided (from cold fusion to the humoral theory of disease).

Even when relative lack of sophistication seems to coincide with gullibility, there is no evidence to suggest the former is causing the latter. On some measures, young children can be said to be more gullible than their older peers or than adults.[53] For instance, it is difficult for three-year-olds to understand that someone is lying to them and to stop trusting them.[54] (In other respects, obviously, three-year-olds are incredibly pigheaded, as any parent who has tried to get their toddler to eat broccoli or go to bed early knows). But this apparent (and partial) gullibility isn't caused by a lack of cognitive maturation. Instead, it reflects the realities of toddlers' environment: compared with adults, small children know very little, and they can usually trust what the adults around them say.[55] In the environment in which we evolved, young children were nearly always in the vicinity of their mothers, who have limited incentive to deceive them, and who would have prevented most abuse. This strong assumption of trustworthiness adopted by young children is, in some ways, similar to that found in bees, which have even fewer reasons to mistrust other bees than young children have to mistrust their caregivers. In neither case does lack of sophistication play any explanatory role in why some agents trust or do not trust others.

The logic of evolution makes it essentially impossible for gullibility to be a stable trait. Gullible individuals would be taken advantage of until they stop paying attention to messages. Instead, humans have to be vigilant. An arms race view of the evolution of vigilance is intuitively appealing, with senders evolving to manipulate receivers, and receivers evolving to ward off these attempts. Even though this arms race view parallels nicely the popular association between lack of sophistication and gullibility, it is mistaken. Instead, openness and vigilance evolved hand in hand as human communication became increasingly broad and powerful. We can now explore in more detail the cognitive mechanisms that allow us to be both open and vigilant toward communication: How do we decide what to believe, who knows best, who to trust, and what to feel.

4

WHAT TO BELIEVE?

IMAGINE YOU ARE A FOODIE. You love all sorts of different cuisines. There's one exception, though: Swiss cuisine. Based on a number of experiences, you have come to think it is mediocre at best. Then your friend Jacques tell you that a new Swiss restaurant has opened in the neighborhood, and that it is really good. What do you do?

Even such a mundane piece of communication illustrates the variety of cues that you ought to consider when evaluating any message. Has Jacques been to the restaurant, or has he just heard about it? Does he particularly like Swiss cuisine, or is he knowledgeable about food in general? Does he have shares in this new venture? The next two chapters are devoted to identifying and understanding the cues that relate to the source of the message. Here I focus on the content of the message.

Imagine that Jacques is as knowledgeable about eating out as you are and has no reason to oversell this restaurant. How do you integrate his point of view—that the Swiss restaurant is great—with your skepticism toward Swiss cuisine? Evaluating messages in light of our preexisting beliefs is the task of the most basic open vigilance mechanism: *plausibility checking*.

On the one hand, it is obvious enough that we should use our preexisting views and knowledge when evaluating what we're told. If someone tells you the moon is made of cheese, some skepticism is called for. If you have consistently had positive interactions with Juanita over the years, and someone tells you she has been a complete jerk to them, you should treat that piece of information with caution.

On the other hand, doesn't relying on our preexisting beliefs open the door to bias? If we reject everything that conflicts with our preexisting views, don't we become hopelessly stubborn and prejudiced?

HOW TO DEAL WITH CONTRARY OPINIONS

Experimental evidence suggests the risk of irrational stubbornness is real. In some circumstances, people seem to become even more entrenched in their views when presented with contrary evidence—to use the earlier example, it is as if you would become even more sure that Swiss cuisine sucks after being told that a Swiss restaurant was great. Psychologists call this phenomenon the *backfire effect*. It has been repeatedly observed; for instance, in an experiment that took place in the years following the second Iraq War. U.S. president George W. Bush and his government had invoked as a reason for invading Iraq the supposed development of weapons of mass destruction by Iraqi leader Saddam Hussein. Even though no such weapons were ever found, the belief they existed persisted for years, especially among conservatives, who had been more likely to support Bush and the Iraq War. In this context, political scientists Brendan Nyhan and Jason Reifler presented American conservatives with authoritative information about the absence of weapons of mass destruction in Iraq.[1] Instead of changing their minds in light of this new information, even a little bit, the participants became more convinced that there had been weapons of mass destruction. A few years later, the same researchers would observe a similar effect among staunch opponents of vaccination: presenting antivaxxers with information on the safety and usefulness of the flu vaccine lowered even further their intention of getting the flu shot.[2]

Surely, though, the backfire effect has to be the exception rather than the rule. Imagine you're asked to guess the length of the Nile. You think it is about seven thousand kilometers long. Someone says it is closer to five thousand kilometers. If the backfire effect were the rule, after several more iterations of the argument, you would be saying the Nile is long enough to circle the earth several times over. Fortunately, that doesn't happen. In this kind of situation—you think the Nile is seven thousand kilometers long, someone else thinks it is five thousand—people move about a third of the way *toward* the other opinion and very rarely away from it.[3]

Even on sensitive issues, such as politics or health, backfire effects are very rare. Nyhan and Reifler had shown that conservatives told about the absence of weapons of mass destruction in Iraq had become even more convinced of the weapons' existence. Political scientists Thomas Wood and Ethan Porter recently attempted to replicate this finding. They succeeded but found this was the only instance of a backfire effect out of thirty persuasion attempts. In the twenty-nine other cases, in which participants were provided with a factual statement relating to U.S. politics (for example, that gun violence has declined, or that there are fewer abortions than ever), their opinions moved in line with the new, reliable information. This was true even when the information went against their preexisting opinion and their political stance.[4] As a rule, when people are presented with messages from credible sources that challenge their views, they move some of the way toward incorporating this new information into their worldview.[5]

In the examples we've seen so far, there was a direct clash between people's beliefs (that there were weapons of mass destruction in Iraq, say) and what they were told (that there were no such weapons). The case of the Swiss restaurant is a little subtler. You don't have an opinion about the specific restaurant Jacques recommended, only a prejudice against Swiss cuisine in general. In this case, the best thing to do is somewhat counterintuitive. On the one hand, you are justified in doubting Jacques's opinion and thinking this new Swiss restaurant is likely to be bad. But you shouldn't then become even more sure that Swiss cuisine in general is poor—that would be a backfire effect. Instead, your beliefs about Swiss cuisine in general should become somewhat less negative, so that if enough (competent and trustworthy) people tell you that Swiss restaurants are great, you end up changing your mind.[6]

BEYOND PLAUSIBILITY CHECKING: ARGUMENTATION

Plausibility checking is an ever-present filter, weighing on whether messages are accepted or rejected. On the whole, this filtering role is mostly negative. If plausibility checking lets in only messages that fit with our prior beliefs, not much change of mind is going to occur—since we already essentially agree with the message. This is why you often need to recognize the qualities of a source of information—that it is reliable and of goodwill—to change your mind. There is, however, an exception, a case in which plausibility checking on its own, with no information whatsoever about the source, gives us a reason to accept a novel piece of information: when the new information increases the coherence of our beliefs.[7]

Insight problems are good examples of how a new piece of information can be accepted purely based on its content. Take the following problem:

> Ciara and Saoirse were born on the same day of the same month of the same year to the same mother and the same father yet they are not twins.
> How is this possible?

If you don't already know the answer, give it a minute or two. Now imagine someone saying "They're part of triplets." Even if you have no trust whatsoever in the individual telling you this, and even though this is a new piece of information, you will accept the answer. It just makes sense: by resolving the inconsistency between two girls being born at the same time of the same mother and their not being twins, it makes your beliefs more coherent.

In some cases, simply being told something is not enough to change our minds, even though accepting the information would make our beliefs more coherent. Take the following problem:

> Paul is looking at Linda.
> Linda is looking at John.
> Paul is married but John is not.
> Is a person who is married looking at a person who is not married?
>
> Yes / No / Cannot be determined.

Think about it for as long as you like (it is one of my favorite logical puzzles, which my colleagues and I have used in many experiments).[8]

Now that you've settled on an answer, imagine that your friend Chetana tells you, "The correct answer is *Yes*." Unless you happen to already think *Yes* is the correct answer, you will likely believe Chetana has gotten the problem completely wrong. You probably reached the conclusion that the correct answer is *Cannot be determined*—indeed, you are likely quite sure this is the correct answer.[9]

Yet Chetana would be right, and you would be better off accepting *Yes* as a correct answer. Why? Because Linda must be either married or not married. If she is married, then it is true that someone who is married (Linda) is looking at someone who is not married (John). But if she is not married, it is also true that someone who is married (Paul) is looking at someone who is not married (Linda). Because it is always true that someone who is married is looking at someone who is not married, the correct answer is *Yes*.

Once you accept the *Yes* answer, you are better off. Yet, among the people who initially provide the wrong answer (so, the vast majority of people), essentially no one accepts the answer *Yes* if they are just told so without the accompanying argument.[10] They need the reason to help them connect the dots.

Arguments aren't only useful for logical problems; they are also omnipresent in everyday life. Going to visit a client with a colleague, you plan on taking the metro line 6 to reach your destination. She suggests taking the bus instead. You point out that the metro would be faster, but she reminds you the metro conductors are on strike, convincing you to take the bus. If you had not accepted her argument, you would have gone to the metro, found it closed, and wasted valuable time.

The cognitive mechanism people rely on to evaluate arguments can be called *reasoning*. Reasoning gives you intuitions about the quality of arguments. When you hear the argument for the *Yes* answer, or for why you should take the bus, reasoning tells you that these are good arguments that warrant changing your mind. The same mechanism is used when we attempt to convince others, as we consider potential arguments with which to reach that end.[11]

Reasoning works in a way that is very similar to plausibility checking. Plausibility checking uses our preexisting beliefs to evaluate what we're told. Reasoning uses our preexisting inferential mechanisms instead. The argument that you shouldn't take the metro because the conductors are on strike works because you naturally draw inferences between "The conductors are on strike" and "The metro will be closed" to "We can't take the metro."[12] If you had thought of the strike yourself, you would have drawn the same inference and accepted the same conclusion: your colleague was just helping you connect the dots.

In the metro case, the dots are very easy to connect, and you might have done so without help. In other instances, however, the task is much harder, as in the problem with Linda, Paul, and John. A new mathematical proof connects dots in a way that is entirely novel and very hard to reach, yet the people who understand the proof only need their preexisting intuitions about the validity of each step to evaluate it.

This view of reasoning helps explain the debates surrounding the Socratic method. In Plato's *Meno*, Socrates walks a young slave through a demonstration of the Pythagorean theorem. Socrates doesn't have to force any conclusion on the slave: once each premise is presented in a proper context, the slave can draw the appropriate conclusions himself. Socrates only has to organize the steps so that the slave can climb them on his own. In a way, the answer has been "given out [by the slave] of his own head,"[13] and yet the boy would likely never have reached it without help.

This illustrates the efficiency of reasoning as a mechanism of open vigilance. Reasoning is vigilant because it prompts us to accept challenging conclusions only when the arguments resonate with our preexisting inferential mechanisms. Like plausibility checking, reasoning is essentially foolproof. Typically, you receive arguments when someone wants to convince you of something you wouldn't have accepted otherwise.[14] If you're too distracted to pay attention to the arguments, you simply won't change your mind. If, even though you're paying attention, you don't understand the arguments, you won't change your mind either. It's only if you understand the arguments, evaluating them in the process, that you might be convinced.

Reasoning not only makes us vigilant but also open-minded, as it helps us accept conclusions we would never have believed without argument. I mentioned previously studies showing that people tend to put more weight on their own views than on other people's opinion, moving only about a third of the way toward the other on average (the length of the Nile example). When people are provided with a chance to discuss the issue together, to exchange arguments in support of their views, they become much better at discriminating between the opinions they should reject and those they should accept—including opinions they would never have accepted without arguments.[15]

The vast majority of our intuitions are sound—otherwise we wouldn't be able to navigate our environment, and we would have been selected out a long time ago. Because we recruit these intuitions to evaluate the reasons people offer us, the reasons we recognize as good enough to warrant changing our minds should, more often than not, lead to more accurate opinions and better decisions. The exchange of arguments in small discussion groups should thus tend to improve performance in a variety of tasks, as people figure out when to change their minds, and which new ideas to adopt. This is exactly what has been observed, as exchanging reasons allows forecasters to make better predictions, doctors to make better diagnoses, jurists to make better judicial decisions, scientists to develop better hypotheses, pupils to better understand what they're taught, and so on.[16]

CHALLENGING ARGUMENTS

To make us more open-minded, reasoning should evaluate arguments as objectively as possible. In particular, people should be able to spot a good argument even if its conclusion is deemed implausible. When we're told that the correct answer to the Linda, Paul, and John problem is *Yes*, plausibility checking says no (for people who got the problem wrong). When your colleague suggests taking the bus, which you know to be slower, plausibility checking says no. Yet in both cases, once the arguments have been presented, reasoning does its job, overcoming the initial rejection suggested by plausibility checking. But maybe you would have been less likely to accept the arguments if you had felt more strongly that their conclusion was wrong?

In an experiment with participants solving the Linda, Paul, and John problem, we got participants so confident in the wrong answer that we had to add some extra options on the confidence scale; otherwise, they all claimed to be "very confident" they had chosen the right answer. Even with the extra options, we ended up with many participants saying they were "as confident as in the things I'm most confident about" that the wrong answer was correct. Yet these ultraconfident participants were just as likely to recognize the right argument, when it was provided to them, as those who were less confident.[17]

Arguments might change people's minds when they relate to riddles. For all their overconfidence, people aren't particularly attached to their (wrong) answers. But what about things that matter: our personal life, politics, religion? Are we still able to evaluate arguments objectively then? Three strands of evidence—experimental, historical, and introspective—make me hopeful that good arguments generally change people's minds, even when they challenge deeply held beliefs.

In many experiments, participants have been offered arguments that vary in strength—from the downright fallacious to the unimpeachable—and asked to evaluate them. In other studies, researchers have measured how much the participants changed their minds as a function of the quality of the arguments they were presented with. The conclusion of these studies is that most participants react rationally to arguments, rejecting fallacious arguments outright, being more convinced by strong than by weak arguments, and changing their minds accordingly.[18]

Historical evidence also suggests that arguments work, even when they support conclusions that are quite revolutionary. In the early twentieth century, some of the West's greatest minds—such as Bertrand Russell, Alfred North Whitehead, and David Hilbert—were attempting to provide a logical foundation for mathematics. In 1930, a young, unknown mathematician named Kurt Gödel offered a proof that this could not be done (more precisely, that it is impossible to find a complete and consistent set of axioms for all of mathematics).[19] As soon as they read the proof, everyone who mattered accepted it, even if that meant jettisoning decades of work and kissing their dream good-bye.[20] Beyond mathematics and its perfect demonstrations, good arguments also work in science, even if they challenge established theories. It is simply not true that, as Max Planck complained, "a new scientific truth does not triumph by convincing its opponents and making them see the light, but rather because its opponents eventually die, and a new generation grows up that is familiar with it."[21] When the evidence is there, new theories are promptly accepted by the scientific community, however revolutionary they may be. For example, once sufficient evidence had been gathered to support plate tectonics, it only took a few years to turn it from fringe theory to textbook material.[22]

Good arguments even work in the political and moral realm. In *The Enigma of Reason*, Dan Sperber and I reviewed the examples from mathematics and science just mentioned but also the amazing story of British abolitionism, when a nation was persuaded to abandon the slave trade in spite of the economic costs involved.[23] In many countries, the past decades have seen dramatic improvements in the rights of women, LGBT+ individuals, and racial minorities. In each of these cases, community leaders, intellectuals, journalists, academics, and politicians spent time and effort developing arguments drawing on a variety of moral and evidential sources. People read and heard these arguments and appropriated some of them in their everyday conversations.[24] Even if good arguments are not the only cause of the momentous changes we have witnessed, the efforts spent developing, laying out, and communicating good reasons for people to change their minds have likely contributed to dramatic changes in public opinion.

On a personal level, I believe we have all experienced, at some time or other, the pull of uncomfortable arguments. When I was going to university, being a (staunch) leftist was very much taken for granted. But I kept encountering arguments that challenged some of the political tenets widely accepted among my peers. Ignoring these arguments would have had no negative practical consequences for me personally—then as now, I have essentially no political power—and would only have brought social benefits in the form of approval by my peers. Yet I couldn't help but feel the strength of these alternative arguments. Even if their conclusions remained for some time somewhat disturbing to me, they played a significant role in shaping my current political opinions.

Arguably, it is this power of a well-thought-out argument that led Martin Luther to develop a distaste—to put it mildly—for reason, distaste that he expressed quite floridly: "Reason is by nature a harmful whore. But she shall not harm me, if only I resist her. Ah, but she is so comely and glittering. . . . See to it that you hold reason in check and do not follow her beautiful cogitations."[25] In the context of the religious battles he was fighting, we can imagine Luther encountering arguments that challenged his moral and religious views. Had Luther been able to easily reject these arguments, had he found no strength in them at all, he surely would not have felt this internal turmoil and developed such resentment toward reason.

WHAT IF OUR INTUITIONS ARE WRONG?

When it comes to evaluating the content of communicated information, people rely on two main mechanisms: plausibility checking, which compares the content of the message with our preexisting beliefs, and reasoning, which checks whether arguments offered in support of the message resonate with our preexisting inferential mechanisms.

I have made the case that plausibility checking and reasoning function well as devices of open vigilance. We can rely on our prior beliefs to evaluate what we're told without falling for a confirmation bias or becoming more polarized because of backfire effects. Reasoning evolved to allow a greater degree of open-mindedness, as good arguments help people accept conclusions they would never have accepted otherwise, even conclusions that challenge deeply held beliefs.

The main issue with plausibility checking and reasoning isn't their potential biases in relating communicated information to our prior beliefs and inferences but the prior beliefs and inferences themselves. Our minds evolved so that most of our beliefs would be accurate, and most of our inferences sound. Indeed, in the wide variety of domains we evolved to deal with—from figuring out what food to eat to what people mean when they speak—we perform well. We also draw sound inferences in domains that are evolutionarily novel, if we've had a lot of opportunities for learning: billions of people are now able to read with near-perfect fluency; human computers—before they were replaced by electronic versions—could perform complex mental calculations with few slipups.

By contrast, if we attempt to draw inferences in any domain that evolution and learning haven't equipped us to deal with, we are likely to be systematically wrong. When tackling novel questions, we grope for solutions and reach for some adjacent cognitive mechanism that appears relevant to the question at hand. This adjacent cognitive mechanism might well be the same for most of those struggling with the same question. When many people get things wrong in the same way, cultural patterns can emerge.

Imagine someone with no access to science wondering why animals have traits that are so well adapted to their environment. We're not equipped with mechanisms to answer this question directly (why would we? the practical benefits are essentially nil). By contrast, recognizing and understanding artifacts matters a lot, and we're likely equipped with cognitive mechanisms to deal with that problem. Artifacts, in their own way, are also adapted to their environment. Because we know that artifacts are made by agents, it appears to make sense that adaptive traits in animals have also been created by agents.[26] This is one of the reasons why creationism has a field day, as people find it intuitively compelling—certainly more compelling than the Darwinian idea that adaptations appear through an unguided process of natural selection.

The same logic applies to many popular misconceptions. Take vaccination. In the popular imagination, it consists in taking healthy babies and injecting them with something that contains a bit of disease (in fact, the agents are typically inert). All our intuitions about pathogens and contagion scream folly.[27] The recent rise in anti-vaccination sentiment has often been blamed on specific people, master persuaders—from Andrew Wakefield in the United Kingdom to Jenny McCarthy in the United States—who would have swayed significant swaths of the population into harmful, scientifically illiterate behavior. In fact, the anti-vaccination movement is as old as vaccination. Already in 1853, England's enactment of the first Compulsory Vaccination Act "provoked enormous fears of contamination."[28] After a lull in the early twentieth century—likely due, at least in part, to the plainly visible success of the polio vaccine—fears about vaccination have rebounded in the West. These fears create a demand for anti-vaccination rhetoric, demand that is rapidly met. As historian of medicine Elena Conis argued, "Both the Wakefield study [linking, falsely, vaccines and autism] and McCarthy's prominence as a vaccine skeptic [are] the products—not the cause—of today's parental vaccine worries."[29] The Wakefield study, in particular, only had an impact on the U.S. rate of MMR vaccination (the vaccine that had been fraudulently linked to autism) around the year 2000, when it was known only by professionals, before it was found to be fake and its results contradicted by dozens of other studies. By contrast, the media frenzy that surrounded the vaccines–autism link, starting a few years later, had no effect on vaccination rates.[30]

The examples can be multiplied. We're not well equipped to think about the politics or economics of large, complex, diverse states. Instead, we have recourse to intuitions that evolved by dealing with small coalitions in conflict with each other.[31] These intuitions tell us, for instance, that if someone else benefits too much from trading with us, then we must be losing out, or that we should be wary of powerful enemy coalitions colluding against us. Hence the popular success of protectionist or, more generally, antitrade policies, and of conspiracy theories (obviously, these intuitions are sometimes right: some trade is bad for us, and some people do conspire).

Not all misconceptions can be neatly explained by the idea of misapplied intuitions. The French are big on homeopathy, and I'm still stumped as to why—how is duck liver diluted up to the point that there's nothing left of it supposed to cure the flu? In chapter 8 I will mention some decidedly nonorthodox beliefs shared by the inhabitants of Montaillou in thirteenth-century France—that preserved umbilical cords could help win lawsuits, for example. They also baffle me (seriously, if you have suggestions, let me know).

Still, compared with, say, creationism, resistance to vaccination, or conspiracy theories, these odd misconceptions are much more culturally specific (the umbilical cord thing in particular hasn't caught on well). Some explanation for their existence is certainly required, but the point stands that popular misconceptions, as a rule, are intuitively compelling. In the absence of strong countervailing forces, it doesn't take much persuasion to turn someone into a creationist anti-vaxxer conspiracy theorist.

Far from being due to widespread credulity, the prevalence of intuitive misconceptions reflects the operation of plausibility checking, when it happens to work with poor material. When more accurate views—in evolution by natural selection, in the efficacy of vaccines, and so forth—spread, it is in part thanks to argumentation, but argumentation is most powerful among people who can discuss issues at length and share a lot of common ground. For these sound beliefs to expend outside of a circle of experts, we must be able to recognize that sometimes others know better.

5

WHO KNOWS BEST?

ON JANUARY 5, 2013, Sabine Moreau, resident of the small Belgian town of Erquelinnes, was supposed to pick up a friend at the train station in Brussels, fifty miles away. She punched the address in her satnav and started driving. Two days and eight hundred miles later she had reached Zagreb, on the other side of Europe, having crossed three countries on the way. That was when she decided that something must be wrong, made a U-turn, and found her way back to Erquelinnes.[1]

As I argued in the last chapter, we put more weight on our own beliefs than on communicated information—when everything else is equal. Everything else often isn't equal. Others can be ignorant, mistaken, or poorly informed, giving us reasons to discount their opinions. But others can also be more competent and better informed. A major part of the open-mindedness of open vigilance mechanisms comes from being able to identify, and then listen to, those people who know better, overcoming our initial reaction driven by plausibility checking to reject information that conflicts with our prior beliefs.

In this chapter, I explore a variety of cues that help us identify who knows best: Who has had the best access to information? Who has a history of getting things right? Whose opinion is shared by the most people?

These cues told Sabine Moreau she should believe her satnav. The satnav has access to precise maps, had proven reliable over many past trips, and is widely believed to be accurate. Obviously, she went too far in letting these cues trump her intuitions. But for one Sabine Moreau, how many people end up lost or stuck in traffic jams because they didn't follow their satnav's suggestions?

EYEWITNESS ADVANTAGE

The most obvious cue that someone else is more likely than us to be right is access to a sound source of information. You believe that your friend Paula is not pregnant. Bill, who you know has just seen Paula, tells you not only that she is but also that she is quite far along in her pregnancy. Assuming you have no reason to think Bill is lying to you (a question we turn to in the next chapter), you should change your mind and accept that Paula is pregnant. Testimony from the right source can also amount to privileged access—if you know Bill has just called Paula, you should also believe him when he tells you she is pregnant.

Intuitions about the value of informational access develop very early. In a classic study, psychologist Elizabeth Robinson and her colleagues systematically varied the information children—some as young as three—had about what was in a box.[2] Some children saw what was in the box; others simply guessed. The children were then confronted by an adult who told them that what was in the box was different from what the children had just said. Like the children, some of the adults had seen what was in the box, and others had just guessed. The children were most likely to believe the adult when she had seen what was in the box, while they had only guessed; the children were least likely to believe the adult when she had guessed what was in the box, while they had seen it.

If we do not already know what information our interlocutors have had access to, they often tell us. If Bill knows you think Paula isn't pregnant, he might preempt your doubts by saying, "Paula just told me she's pregnant." Such information about the source of our beliefs is ubiquitous in conversation. Even when none is mentioned explicitly, some can usually be inferred. If Bill tells you "That movie is great!," this suggests he saw the movie rather than, say, read some reviews.

Again, even young children are sensitive to such reported informational access. In a series of experiments, my colleagues and I asked preschoolers to help a (toy) girl find her lost (toy) dog. A (toy) woman suggested the dog had gone in one direction, saying that she had seen it go that way. Another (toy) woman pointed to a different direction, without specifying why she believed this to be the right place to look for the dog. The children were more likely to believe the woman mentioning a reliable access to information, and the same was true when the second woman provided a bad reason rather than no reason at all.[3]

RELIABLE EXPERTISE

When a friend offers a way of fixing our computer problems, recommends a restaurant, or provides dating advice, it is not enough to know where she got her ideas from. Maybe she had firsthand experience of the restaurant, but the value of that experience depends on her taste in food—if she can't tell a McDonald's from a Michelin-starred restaurant, her firsthand experience isn't worth much. How can we figure out who is competent in what domain?

The most reliable cue is past performance. If someone has been consistently able to solve computer problems, pick exquisite restaurants, or give sound dating advice, then it is probably worth listening to what they're saying in each of these domains.

From an evolutionary point of view, what makes past performance a great cue is that it is hard or impossible to fake. It is difficult to consistently solve computer problems, find exquisite restaurants, and give sound dating advice if you don't possess some underlying skill or knowledge that allows you to keep performing well in these areas.

To evaluate others' performance, we can rely on a wide variety of cognitive devices. Humans are equipped with mechanisms for understanding what other people want, believe, and intend. Thanks to these mind-reading mechanisms, we can understand, say, that our friend wants her computer to work again. All we have to do, then, is keep track of whether she successfully reaches her goal.

We can also rely on the mechanisms described in the last chapter: plausibility checking and reasoning. Someone who gives you the right answer to an insight problem (like the one with the triplets) or who offers a novel and convincing mathematical demonstration should be deemed more competent, at least in these domains.[4]

Seeing someone, from a professional athlete to a craftsperson, do something well can be very pleasurable, giving rise to so-called competence porn—for example, the consistently articulate and witty exchanges of screenwriter Aaron Sorkin's protagonists. The pleasure we derive from watching someone perform flawlessly actions that do not benefit us directly is likely related to the learning possibilities afforded.

Having recognized who is good in what area, one possibility is to imitate them. Some nonhuman animals, such as the house mouse, are already selective in who they imitate, being more likely to copy the actions of adults than of juveniles.[5] But imitation has its limits. Copying what your friend does when she fixes her computer problems is unlikely to help fix your own problem. Following your foodie friend around might lead you to some places that are not really to your taste, and to a depleted bank account. That's when communication comes in handy. Once you have inferred, through past performance, that a given friend is good with computers, you can ask them about your specific problem. You can request from your foodie friend recommendations that fit your taste and your budget. Drawing on your friends' expertise to get answers to your own problems makes more sense than simply imitating them.[6]

EINSTEIN OR THE MECHANIC?

Looking at past performance is a powerful strategy for establishing competence, but it is not as simple as it seems. One difficulty is that performance can be largely a matter of luck. A notorious contemporary example is the trading of stocks in financial markets: it is fantastically difficult to tell whether the performance of, say, a hedge fund is due to the intrinsic competence of its traders or to dumb luck.[7] Even strong performance over several years is not much of an indicator: given how many hedge funds there are, it is statistically unavoidable that some will perform well year after year through chance alone. The same logic applies to skills that were undoubtedly more relevant during earlier stages of human evolution, such as hunting large game. Once a given level of competence is reached, who makes the kill on any given day is partly a matter of luck, making it difficult to assess the hunters' individual competence.[8]

Fortunately, performance in many domains—fixing computers, say—is less erratic.[9] But even when performance can reliably be detected, an issue persists: How do we generalize from observed performance to underlying competence? When your friend fixes the printer problem on her computer, what should you infer? We intuitively discard some options: that she is good at fixing things on Mondays, at fixing gray things, or at fixing things on tables. But that leaves a wide array of plausible possibilities: that she's good at fixing printer problems on her computer, fixing problems on her computer, fixing printer problems, fixing Macs, fixing computers, fixing electronics, fixing things, understanding complex problems, or following instructions.

The fact is that psychologists do not know how people *should* generalize from observed performance to underlying competence. Some psychologists argue that competence on many cognitive tasks is related through IQ. Others think we have different kinds of intelligences. Robert Sternberg, for example, has developed a theory of three aspects of intelligence: analytic skills, creativity, and practical skills. Howard Gardner argues for eight or nine modalities of intelligence, ranging from visual-spatial to bodily-kinesthetic.[10] Other psychologists yet suggest our minds are made up of a great many specialized modules—from a face-recognition module to a reasoning module—and that the strength of these modules varies from one individual to the next.[11]

Whatever the correct answer to this complex problem turns out to be, it is clear that humans are endowed with intuitions to guide their inferences from performance to underlying competence. These intuitions are already on display in young children. Preschoolers know that they should direct questions about toys to another child rather than an adult, and questions about food to an adult rather than a child.[12] When they are asked who knows more about how elevators work, preschoolers pick a mechanic over a doctor, while they pick a doctor over a mechanic when asked who knows more about why plants need sunlight to grow.[13]

Adults also appear to be quite good at telling who is good at what. As we saw earlier, intraindividual variability in hunting performance means that sustained observation over long periods of time is necessary to tell who is the best hunter. Yet people are capable of such observations. When Hadza—a group of traditional hunter-gatherers from southern Africa—were asked to evaluate the hunters in their community, their rankings correlated well with hunting performance (as measured by the experimenters, for instance, by testing archery skills).[14]

Moving from the plains of Tanzania to the pubs of South West England, a recent experiment asked groups of participants from Cornwall a series of quiz questions on a wide range of trivia problems, from geography to art history.[15] Participants were then asked to nominate a group member to answer, on their own, some bonus questions—if they answered well, the whole group would benefit. Even though the participants hadn't received any feedback on who had given the correct answers in the initial round of questions, they didn't rely on rough heuristics, such as picking dominant or prestigious individuals. Instead, they were able to accurately select the most competent members in each trivia category. More significantly, research on political discussions shows that U.S. citizens are able to figure out who, among the people they know, is most knowledgeable about politics—and they are more likely to broach political topics with these more knowledgeable acquaintances.[16]

RATIONAL SHEEP

That a given individual knows more than we do, either because they have had access to better information or because they are more competent, is not the only cue telling us that other people might be right and we might be wrong (or simply ignorant). To evaluate an opinion we can look beyond the individual competence of whoever holds it and take into account how many people hold it.

Accepting something because it is the majority opinion has a bad press. For millennia people have been castigated for indiscriminately following the crowd. This distaste for majority opinion has led some intellectuals to pretty extreme conclusions, as when philosopher Søren Kierkegaard claimed that "truth always rests with the minority,"[17] or Mark Twain concurred that "the Majority is always in the wrong."[18]

By this logic, the earth is flat and ruled by shape-shifting lizards. Without being as pessimistic as Kierkegaard or Twain, some experiments suggest that people put little stock in majority opinion. Take the following quiz:

> Imagine an assembly that contains ninety-nine members. They have to decide between two options: option 1 and option 2. One option is better than the other, but, before the vote, we don't know which.
>
> To decide between the two options, they use majority voting. The ninety-nine members vote, and if one option gathers fifty votes or more, then it wins.
>
> Each member of the assembly has a 65 percent chance of selecting the best option.
>
> What do you think are the odds that the assembly selects the best option?

WHO KNOWS BEST? 71

Martin Dockendorff, Melissa Schwartzberg, and I put this question and variants of it to participants in the United States.[19] On average the participants believed that the assembly was barely more likely than chance to select the best option. Majority voting would thus make the assembly no more likely to select the best option than each individual member—quite an indictment of democratic procedures.

In fact, there is a correct answer to this question. Its formula was discovered in the late eighteenth century by the Marquis de Condorcet,[20] an extraordinary intellectual who defended the French Revolution yet ended up having to kill himself to escape the guillotine. Thanks to the Condorcet jury theorem, we know that the odds of the assembly being right are in fact 98 percent (making a few assumptions to which I will return later).

The power of aggregating information from many sources is increasingly recognized. A century after Condorcet, Francis Galton showed how averaging across many opinions is nearly guaranteed to lower the resulting error: the error of the average is generally lower, and never worse than the average error.[21] Much more recently, journalist James Surowiecki brilliantly popularized the "miracle of aggregation" in *The Wisdom of Crowds*.[22] Cartoonist Randall Munroe made this logic intuitive with his xkcd "Bridge" strip (figure 2).[23]

The potential benefits of following the majority are so substantial that many nonhuman animals rely on this heuristic.[24] Baboons offer a perfect example. They travel in troops of several dozen members. The individuals forming the group must constantly make decisions about where to look for food next. To study their decision making, Ariana Strandburg-Peshkin and her colleagues outfitted the baboons from a troop with a GPS, allowing the researchers to closely track their movements.[25] Sometimes, the group starts splitting, with two subgroups

FIGURE 2. "Bridge" strip from the webcomic *xkcd* by Randall Munroe. *Source:* xkcd.com.

initiating a move in different directions. When this happens, the other baboons look at the numbers in each subgroup: the larger one subgroup is compared with the other, the more likely the baboons are to follow it.

If baboons and other animals living in groups have an intuitive grasp of the power of majority rules, it would be bizarre if humans, who rely much more on social information than baboons, completely neglected this rich source of insight.[26]

The study mentioned earlier with the ninety-nine assembly members, and others like it, show that people do not grasp the power of majority rules when the question is framed abstractly, with the number of people forming the majority being represented by a figure or percentage. Different results are obtained when we can see the opinions of individual people. In one of the cleanest studies on the topic, psychologist Thomas Morgan and his colleagues had participants solve a variety of tasks, for instance, deciding whether two shapes seen from different angles are the same or different, tasks that were built to be difficult enough that participants would not be certain of their own answers.[27] The participants were given the individual answers of a number of other participants (or so they thought; in fact, the answers were made up by the experimenters). In this context, people were perfectly able to follow majority rules rationally: they were more likely to change their minds when a given answer was supported by a larger group, and when it was more consensual (holding group size constant). Such a pattern has been found in many other experiments, and it emerges in children during the preschool years.[28]

We have an intuitive grasp of the power of majority rules. Yet, when the question is framed in more explicit, abstract terms, we display no such comprehension. A way of making sense of this apparent contradiction is to introduce the concept of an *evolutionarily valid cue*.[29] A cue is evolutionarily valid if it was present and reliable in the relevant period of our evolution. For instance, our ancestors were better off staying away from rotting meat; rotting meat generates ammonia; ammonia is an evolutionarily valid cue that food should be avoided; this is why we find its odor so repulsive (think cat pee).

Assuming that other primates, like the baboons, are able to follow majority rule, the number of individuals making a given decision has likely been a reliable cue for a very long time, long before the human line split from that of the chimpanzees. By contrast, numbers, odds, and percentages are recent cultural inventions.[30] For this reason, they do not constitute evolutionarily valid cues. While it is possible to react appropriately to such cues—as when we go through the equation of the Condorcet jury theorem—this takes dedicated, explicit learning.

We find the same pattern when it comes to considering whose opinion influenced whom.[31] The Condorcet jury theorem fully applies if the voters have formed their opinions independently of each other. If ninety-eight members of our imaginary assembly mindlessly follow the ninety-ninth member, the opinion of the assembly is only as good as that of the ninety-ninth member.

Abstract cues related to the dependence between opinions, such as correlation coefficients, are simply ignored.[32] This makes sense given that correlation coefficients are definitively not an evolutionarily valid cue. By contrast, when Helena Miton and I introduced evolutionarily valid cues, participants took them into account.[33] Our participants were told of three friends who had all recommended a given restaurant. However, they had done so because a common fourth friend had told them the restaurant was great. In this case, participants knew to treat the opinion of these three friends as if it were the opinion of a single friend. Other experiments suggest that even four-year-olds are able to take some forms of dependence between opinions into account.[34]

RESISTING THE PULL OF THE MAJORITY

When cues are evolutionarily valid, people weigh a number of them to decide how valuable the majority opinion is: size of the majority in relative terms (the degree of consensus) and absolute terms (group size), competence of the members of the majority, and degree of dependency between their opinions.[35] But how do they weigh all this in relation to their own prior state of mind? As we saw in chapter 1, it is widely believed that when these cues converge, in particular when people are faced with a large consensual group, the force of the majority opinion becomes nearly irresistible.

Asch's conformity experiments, in which the consensus of a dozen individuals had people doubting their own eyes, believing that two lines of clearly different length were just as long, seem to offer the perfect demonstration of the power of majority opinion to quash prior beliefs, however strong they are. But Asch himself never framed his experiments in this way, stressing instead the power of the individual to resist group pressure. After all, the participants followed the crowd in only about one-third of the trials.[36] Even then, it was the social pressure to conform that drove this behavior, not the informational pull of the majority opinion.[37] After the experiment was finished, many participants admitted to having yielded even though they knew the group to be wrong.[38]

An even better proof rests with another version of the experiment in which participants were told that they had arrived late and, as a result, would have to write down their answers on a sheet of paper. Instead of having to voice their opinions in front of the group, which unanimously agreed on an obviously wrong line, they could answer privately. Conformity plummeted. In only a small minority of trials did participants choose to follow the consensus.[39] Across all these experiments, a very small number of people genuinely believed the group to be right, and they relied on a variety of strategies to make sense of the group's weird answer—that the lines created a visual illusion, or that it was their width rather than their length that was the object of the task.[40]

What about, then, Milgram's conformity experiments, in which a few confederates looking up at a random building were seemingly able to get every passerby to do the same? A more recent take on this study shows that, unlike Asch's participants, the passersby were not conforming because of social pressure. Psychologist Andrew Gallup and his colleagues replicated Milgram's experiment, asking a varying number of confederates on busy sidewalks to look up in the same direction.[41] This time, however, the confederates were looking up at a camera and, using motion-tracking software, the researchers could describe in detail the behavior of the people around them. As in Milgram's experiment, some people started looking up as well, but they were more likely to do so when they were behind, rather than in front of, the confederates. This rules out social pressure as a significant force, for then people would have started looking up when they could be seen and judged by the confederates. But Gallup also found that people reacted quite rationally on the whole, using sound cues to decide when to follow majority opinion. The passersby were more likely to look up when more confederates were doing so, and when the crowd was thinner (and so a greater proportion of people were looking up). The researchers also found that many people didn't look up, showing the reaction not to be reflex-like at all. Instead, whether people looked up was likely determined by other factors, such as whether or not they were in a hurry.

COMPETENT COMPETENCE DETECTION

On the whole, our mechanisms of open vigilance provide us with a good estimate of who knows best. Preschoolers already use an impressive number of cues to decide whether others might know more than they do. They consider who has the most reliable informational access. They use past performance to establish who is more competent in a given domain.[42] They have a good idea of the boundaries of a specialist's domain of expertise. They are more likely to follow the majority opinion when the majority is more numerous and closer to a consensus, when its members are competent, and when they have formed their opinions independently of each other.[43]

Preschoolers are open to changing their minds when they think others know better, yet they remain vigilant. Far from blindly following prestigious but incompetent individuals or the majority opinion, they weigh cues to competence and conformity against their prior beliefs, so that even in the face of experts or a strong consensus, they do not automatically change their minds—and neither do adults, contra a naive interpretation of the Asch conformity experiments.

This doesn't mean we cannot be tricked by people who seem competent or who are able to fake a consensus. Fool's errands are an innocuous example. When an apprentice in a workshop is sent to look for elbow grease, all the cues tell him that he should comply. The workers asking him are competent, they all agree, and, if prompted, they provide independent reasons for the importance of the task at hand.[44] To avoid looking like a fool, the apprentice has to ignore these cues and consider that all of these people might not have his best interests at heart.

6

WHO TO TRUST?

IN ORDER TO PROPERLY evaluate what we are being told, we need to figure out who knows best, but this isn't enough. The most competent expert is useless if they decide to lie. A consensual group is of equally little use if its members conspire to deceive us.

A tremendous amount of research has been dedicated to the question of "deception detection," in other words, how good we are at spotting liars. Are we any good at it? What cues do we rely on? How reliable are these cues? The practical stakes seem huge. From human resource personnel to detectives, from betrayed spouses to victims of telephone scams, who wouldn't like a failsafe way to spot a liar?

Subtle nonverbal cues are often thought to be reliable tells. Someone who fidgets, who looks shifty, or who is unwilling to make eye contact is less likely to inspire confidence.[1] For Freud, "No mortal can keep a secret. If his lips are silent, he chatters with his finger-tips; betrayal oozes out of him at every pore."[2]

Indeed, many people are quite confident in their ability to spot a liar. This is one of the reasons why, in many cultures, verbal testimony is preferred to written testimony in court: judges think they can tell if someone is lying by watching them speak in person.[3] To this day, many detectives are taught to rely on "such visual cues as gaze aversion, nonfrontal posture, slouching, and grooming gestures."[4]

The TV show *Lie to Me* rests on this premise. The show's hero, Cal Lightman, is inspired by Paul Ekman, a psychologist famous for his studies of emotional expressions. Like Ekman, Lightman travels to faraway places in order to demonstrate that people make the same face to express, say, fear, everywhere in the world. Again like Ekman, Lightman uses his deep knowledge of emotional expressions to catch liars, in particular by relying on the observation of microexpressions.[5]

Microexpressions are facial expressions that last for the blink of an eye, less than a fifth of a second. These rapid expressions are supposed to betray the conflicting emotions of those who try to lie, or to hide their feelings more generally. People who want to conceal their guilt, their sadness, or their joy might let slip a tiny muscle movement reflecting the emotion they attempt to hide rather than the one they want to display.

Even though they are essentially invisible to the untrained eye, microexpressions would be perceptible by those who receive the proper instructions, such as the short class offered by Ekman to a variety of law enforcement agencies (and which you can easily buy online). It seems we've finally found a solution to catching liars, and it's only a few hours' training away.

MICROEXPRESSIONS, SCHMICROEXPRESSIONS

Unfortunately, things are not quite so simple. Ekman's ideas, and his results, have proven controversial. Ekman's critics point out that his findings about the reliability of microexpressions as a tool for spotting liars have not been published in proper peer-reviewed journals: he has not shared his methods and data with the scientific community, so they cannot be independently evaluated.[6] Moreover, experiments conducted by scientists outside his group have yielded rather negative results.

Psychologists Stephen Porter and Leanne ten Brinke showed participants stimuli designed to elicit a variety of emotions, from disgust to happiness, while asking some of the participants to display a different emotion than the one that would normally be elicited by the stimuli.[7] They asked coders to judge the participants' facial expressions frame by frame (104,550 of them!) in order to detect the slightest slipups. In nearly a third of the trials in which they had to fake an emotion, participants displayed some inconsistent feelings: someone shown a disgusting image might express fear or happiness, however briefly.

If this appears to vindicate Ekman's theories, it really doesn't, for two reasons. First, these slipups lasted at least one second on average, several times longer than microexpressions are supposed to last, making them easily perceptible by untrained observers. Second, out of the fourteen genuine microexpressions, six were displayed when participants were not attempting to hide anything. This makes microexpressions quite useless as a tool for detecting deception attempts. Another study by ten Brinke and her colleagues, in which participants displayed either genuine or fake remorse, yielded similar results: microexpressions were rare, and the participants who were genuinely remorseful were just as likely to display them as those who were faking it.[8]

The problem extends beyond microexpressions. In the original study by Porter and ten Brinke, a third of the deceitful participants—those who had been asked to fake a facial expression—briefly displayed an expression different from the one they were supposed to display. But that was also true for 27 percent of those who were not asked to fake anything. After all, we often feel conflicting emotions.[9] As a result, such slipups are completely unreliable for detecting deception, failing to catch most deceivers, and catching many who have nothing to hide.

Porter and ten Brinke's results fit a pattern. Dozens of studies have looked in detail at people as they tell lies or true stories, scanning every detail of their behavior for cues to deception. Meta-analyses of these experiments yield pretty grim results: no cue is strong enough to reliably tell who lies and who doesn't.[10] For instance, the correlation between amount of eye contact and likelihood of deception is literally 0, while it is a very modest, and essentially useless, .05 for gaze aversion.[11] As stated in a recent review of this literature: "Prominent behavioral scientists consistently express strong skepticism that judgments of credibility can reliably be made on the basis of demeanor (body language) cues."[12] Because no reliable cue exists, even supposed experts, people who are paid to spot liars as a job, are no more likely than chance to tell, from behavioral cues alone, who lies and who doesn't.[13]

HOW NOT TO CATCH A LIAR

Why are there no reliable behavioral cues to lying and deception? As mentioned earlier, a proximal reason—a reason that relies on the functioning of our psychological mechanisms—is that people feel conflicting emotions whether they are lying or telling the truth, which makes it difficult to distinguish liars from truth tellers. An ultimate reason—a reason that relies on evolution—is that such cues couldn't be evolutionarily stable. If such cues had ever existed, they would have been selected out, in the same way that a poker player should not have any tell when bluffing, at least if they want to keep playing poker and remain solvent. Evolutionarily valid behavioral cues to deception would be maladaptive—and, indeed, there don't seem to be any.

You may be wondering if that isn't a major problem for the argument I have been making, namely, that we are naturally vigilant toward communicated information. How can we be vigilant if we can't tell a lie from the truth? To make things worse, in most lie detection experiments, participants tend to err on the side of thinking people are telling the truth.

Some researchers, most notably psychologist Tim Levine, have argued that this behavior makes sense because people actually tell very few lies.[14] Studies of lying in everyday life suggest that lies are rare—fewer than two a day on average—and that most of them are innocuous, such as feigning to be happier than we are (this is at least true among some samples of Americans).[15] Instead of spending a lot of energy catching these minor deceits, we're better off assuming everyone is truthful. This is reminiscent of the argument developed by philosopher Thomas Reid in the eighteenth century, when he claimed that our "disposition to trust in the truthfulness of others, and to believe what they tell us" is related to our "propensity to speak the truth."[16]

From an evolutionary perspective, the Reid/Levine argument doesn't hold. Given how often senders would benefit from lying, if the amount of lying wasn't somehow constrained, it would balloon until no one could be trusted anymore. If we simply assumed people were generally truthful, they would stop being truthful. I'm sure you can think of a few lies you'd tell if you were guaranteed that people would believe you and you'd never get caught.

If we cannot rely on behavioral cues, how do we deal with the issue of deceit in communication? How do we know who to trust?

NEGLIGENCE AND DILIGENCE

Because deceit relies on hidden intentions, it is intrinsically hard to detect. We would have no idea what most people's intentions are if they did not tell us. In many cases concealing one's intention is as simple as not voluntarily disclosing it. This is why proving in court that someone perjured themselves is difficult: it must be established not only that the accused was mistaken but also that they knew the truth and intentionally withheld it.[17]

But deceit is not the only, or even the main, danger of communication.[18] Imagine that you are looking to buy a used car. The salesman might lie to you outright: "I have another buyer very keen on this car!" But he is also likely to give you misguided advice: "This car would be great for you!" He might believe his advice to be sound, and yet it is more likely to be driven by a desire to close the sale than by a deep knowledge of what type of car would best suit your needs. Now you ask him: "Do you know if this car has ever been in an accident?" and he answers "No." If he knows the car has been in an accident, this is bad. But if he has made no effort to find out, even though the dealership bought the car at a suspiciously low price, he is culpable of negligence, and this isn't much better. In the case at hand, whether he actually knew the car had been in a crash, or should have known it likely had been, makes little difference to you. In both cases you end up with a misleading statement and a lemon.

Deceit is cognitively demanding: we have to think of a story, stick to it, keep it internally coherent, and make it consistent with what our interlocutor knows. By contrast, negligence is easy. Negligence is the default. Even if we are equipped with cognitive mechanisms that help us adjust our communication to what others are likely to find relevant, making sure that what we say includes the information our interlocutor wants or needs to hear is difficult. Our minds are, by necessity, egocentric, attuned to our own desires and preferences, likely to take for granted that people know everything we do and agree with us on most things.[19]

We should thus track the relative diligence of our interlocutors, the effort they put into providing information that is valuable to us. Diligence is different from competence. You might have a friend who is very knowledgeable about food, able to discern the subtlest of flavor and to pick the perfect wine pairings. It makes sense to ask her for tips about restaurants. But if she makes no effort whatsoever to adjust her advice to your circumstances—your tastes, your budget, your food restrictions—then the advice is not of much use. If you repeatedly tell her you are a vegetarian and she sends you to a steakhouse, she hasn't been diligent in finding the right information to communicate. You would be justified in resenting this failure, and trusting her advice less in the future.

Stressing diligence—the effort people make in sending us useful information—over intent to deceive shifts the perspective. Instead of looking for cues to deception, that is, reasons to reject a message, we should be looking for cues to diligence, that is, reasons to accept a message.[20] This makes more sense from an open vigilance point of view as the baseline, then, is to reject what we're told, unless some cues suggest our interlocutors have been diligent enough in deciding what to tell us.

INCENTIVES MATTER

When are our interlocutors likely to have done due diligence (including, obviously, not deceiving us)? Simply, when their incentives are aligned with ours: when they're better off if we're better off. There are, broadly, two reasons why incentives between different individuals would be aligned. Sometimes, incentives are naturally aligned. For example, if you and your friend Hadi are moving a clothes dryer together, you both have an incentive to do it as effortlessly as possible by coordinating your actions, so that you lift at the same time, go in the same direction, and so forth. As a result, if Hadi tells you, "Let's lift at three," you have every reason to believe he will also be lifting at three. Other natural alignments in incentives are more long term: parents have a natural incentive for their children to do well, and good friends for each other to be successful.

A simple thought experiment tells us whether or not incentives are naturally aligned: we just have to consider what would happen if the receiver of the information didn't know who the sender really was. For example, Hadi would still want you to know that he will be lifting his end of the dryer at three, even if he weren't the one telling you when he'd be lifting. Likewise, a mother who wants to convince her son that he should study medicine doesn't care if she is the one doing the convincing, as long as the son ends up a doctor.

On the whole, we are quite good at taking natural alignments between incentives into account: when we have evidence that our incentives and those of the sender are well aligned, we take their opinion into account more. This is neatly demonstrated in a study by psychologist Janet Sniezek and her colleagues.[21] Advisers were asked for their opinions on a random topic (the price of backpacks), and the researchers observed how much participants took this opinion into account. After they had received feedback on the real price of the backpacks, some participants could decide to reward the advisers, and the advisers knew this before offering their advice. The advisers had incentives to provide useful advice, and this was mutual knowledge for them and the participants. As a result, participants put more weight on the opinion of these advisers, whose incentives were aligned with theirs.[22]

A much more dramatic example of our ability to start trusting people when we realize our incentives are aligned is that of Max Gendelman and Karl Kirschner.[23] Gendelman was a Jewish American soldier who had been captured by the Germans in 1944 and was held prisoner close to the eastern front. Kirschner was a wounded German soldier recovering at home, close to the prison camp. The two had met while Gendelman was in the camp; when he managed to escape, he took refuge in Kirschner's home. Kirschner told Gendelman that, as a German soldier, he had to flee the Russian advance, and that the two should help each other. Gendelman needed Kirschner to avoid being shot by the Germans looking for escapees. Kirschner needed Gendelman to avoid being shot by the Americans once they would reach their lines. This alignment in their incentives allowed the two former enemies to communicate and collaborate until they were safely behind American lines.[24]

If people can put more weight on what their interlocutor says when they detect an alignment in their incentives, they can also stop listening to what their most trusted friends or dearest family members say when their incentives aren't aligned. This is what happens when friends play competitive games, from poker to Settlers of Catan. Primary-school children are also able to take incentives into consideration when deciding whether to believe what they're told. Psychologists Bolivar Reyes-Jaquez and Catharine Echols presented seven- and nine-year-olds with the following game.[25] Someone—the witness—would see in which of two boxes a candy was hidden. Someone else—the guesser—would pick one of the two boxes to open. The witness could suggest to the guesser which box to open. In the cooperative condition, both witness and guesser got a treat if the guesser opened the right box. In the competitive condition, only the guesser got a treat if the right box was open, and the witness got a treat if the guesser opened the wrong box. In the cooperative condition, children in the role of guesser always believed the witness. By contrast, in the competitive condition, they rightfully ignored what the witness was saying, picking boxes at random.[26]

Following the same logic, children and adults are wary of self-interested claims. Seven-year-olds are more likely to believe someone who says they just lost a contested race than someone who claims to have won it.[27] Plausibly the most important factor adults take into consideration when deciding whether or not someone is lying is that individual's motivation: having an incentive to lie makes someone's credibility drop like a stone.[28]

Incentives can be more or less naturally aligned, but they are rarely, if ever, completely aligned. Hadi might like you to carry the heavier side of the dryer. The mother might want her son to be a doctor in part because she would gain in social status. Your friend might want you to be successful, but perhaps not vastly more successful than himself. Fortunately, humans have developed a great way of making incentives fall in line: reputation.

REPUTATION GAMES

Considering the natural alignment of incentives is essential when assessing the credibility of the sender of a message, but it is not enough on its own, as it does not help us solve the essential problem with the evolution of communication: What happens when incentives diverge?[29]

What we need is some artificial way of aligning the incentives of those who send and those who receive information. Punishment might seem like the way to go. If we punish people who send us unreliable information, by beating them up, say, then they have an incentive to be careful what information they send us. Unfortunately (or not), from an evolutionary perspective this intuitive solution doesn't work as well as it seems. Inflicting punishment on someone is costly: the sender being beaten up is unlikely to take their punishment passively. If the damage from the harmful message is already done, incurring the further cost of punishing its sender doesn't do us any good. Punishment is only valuable as a deterrent: if the sender can be persuaded, before they send a message, that they will be punished if they send an unreliable message, they will be more careful.[30]

The question thus becomes one of communication: How do we communicate that we would be ready to punish people who send us unreliable messages? At this stage, the conundrum of the evolution of reliable communication rears its head. Everybody, including people who don't have the means or the intention of punishing anybody, would be better off telling everyone they will be punished if they send unreliable messages. Individuals who would genuinely punish unreliable senders need a way of making their signals credible. Far from solving the problem of reliable communication, punishment only works if the problem of reliable communication has already been solved.

Fortunately, humans have evolved ways of cooperating and aligning their incentives by monitoring each other's reputations.[31] For a very long time, humans who were either bad at selecting cooperation partners or at being cooperation partners haven't fared very well, at least on average. For the worst cooperation partners, ostracism was like a death sentence: surviving on our own in the wild is next to impossible.[32] As a result, we have become very good at selecting cooperation partners and at maximizing the odds that others would want to cooperate with us.[33]

Being a diligent communicator is a crucial trait of a good cooperation partner. Receivers should be able to keep track of who is diligent and who isn't, and adjust their future behavior on this basis, so that they are less likely to listen to and cooperate with people who haven't been diligent. If this is true, senders have an incentive to be diligent when they communicate with receivers with whom they might want to cooperate—or even receivers who might influence people with whom they might want to cooperate. The incentives between senders and receivers have been socially aligned.

If, thanks to the social alignment of incentives, we can increase senders' willingness to be diligent in what they tell us, we can't expect them to always be maximally diligent. This would be an unfair demand. For Simonetta, a foodie friend of yours, to give you the best possible advice, she would have to learn everything about your tastes, which restaurants you have eaten at lately, what price you can afford, who you plan on taking out, and so on. If people were always expected to be maximally diligent, and if they could expect negative consequences—decreased trust, less cooperation—from failing to reach this goal, they would often end up saying nothing, depriving us of potentially valuable information. After all, a tip from your foodie friend might still be useful, even if she hasn't taken every potential factor into account in formulating her advice.

What we need is a way of managing expectations, a way for senders to tell receivers how much weight to put on the messages. This is the function of commitment signals.[34] We can indicate how much we commit to everything we say (or write). When we commit, we are essentially telling receivers that we are quite sure the message will be valuable to them. Audiences should be more likely to accept the message, but they should also react more strongly if it turns out we haven't been very diligent after all.

Commitment signals abound in human communication. Some commitment markers are explicit indications of confidence: "I'm sure," "I guess," "I think," and so forth. Epistemic modals—*might*, *may*, *could*—also modulate commitment. Increased confidence (and thus commitment) is also signaled implicitly through nonlinguistic signals such as a greater variation in pitch.[35] Even the sources we provide for our beliefs have implications for our degree of commitment. Saying "I've seen that Paula is pregnant" commits us more to the truth of Paula being pregnant than does "People say Paula is pregnant." Even young children are able to take these signals into account: for example, two-year-olds are more likely to believe confident speakers.[36]

Adjusting how much we take into account what people say as a function of their level of commitment is sensible, but only if two conditions are met: not everybody's commitments should be treated equally, and we should adjust our future trust in light of past commitment violations.

ELEPHANTS DO NOT FORGET (WHEN PEOPLE HAVE BEEN OVERCONFIDENT)

To properly take commitment into account, we must keep track of how much our interlocutors value our continued cooperation, so we know how much weight to put on their commitment. The more we think they want to keep cooperating with us, the more we can trust their commitment. But we must also keep track of who is committed to what so that we can adjust our trust accordingly. If commitment signals could be used wantonly, without fear of repercussions, they would not be stable. In an attempt to influence others, everybody would constantly commit as much as possible, making the signals worthless.

In a series of experiments, psychologist Elizabeth Tenney and her colleagues asked participants to deliver a verdict in a mock trial on the basis of two testimonies, delivered by two witnesses, one of which was more confident than the other.[37] Because participants had no other way of distinguishing between the testimonies, they put more faith in the more confident witness. Later on, both witnesses were revealed to have been wrong, and participants now found the witness who had been less confident to be more credible: she had been as wrong as the other witness, but less committed.

Colleagues and I have replicated and extended these findings.[38] Participants were exposed to two advisers who provided the same advice but with different degrees of confidence. After it was revealed that both advisers had been wrong, participants were more willing to trust, even in a completely unrelated domain, the adviser who had been less confidently wrong.

Saying that overconfidence doesn't pay might seem odd. Aren't there many successful politicians or businesspeople who are full of bluster? It is possible that overconfidence might pay in some settings, as when we lack good feedback on the speaker's actual performance. However, it is important to keep in mind that these settings are the exception rather than the rule. In small groups, when people can monitor each other's words and deeds quite easily, overconfidence is a poor strategy. Small-scale traditional societies, which are relatively close to the environment our ancestors lived in, offer a great example. When an individual becomes a leader in such a society, it isn't thanks to empty bluster but because they have superior practical skills, give better advice, or know how to resolve conflicts.[39] We also see this in our everyday lives. Some of us might have been taken in by an ebullient friend who swears their ideas are the best. But this is short-lived. If their ideas don't pan out, we learn to adjust our expectations.

We are more influenced—everything else being equal—by more committed speakers, those who express themselves more confidently. But those more committed speakers lose more if it turns out they were wrong. These costs in terms of lost reputation and ability to influence others are what keep commitment signals stable.

DECIDING WHO TO TRUST

Deciding who to trust has little to do with looking for signs of nervousness or attempting to spot elusive microexpressions. It is not even primarily about catching liars. Instead, it is about recognizing who is diligent in communicating with us: who makes the effort to provide information that is useful to us, rather than only to them. And diligence is all about incentives: we can trust speakers to be diligent when their incentives align with ours.

Incentives between sender and receiver are sometimes naturally aligned—when people are in the same boat. However, even minor misalignment in incentives can create communication breakdowns, so that naturally aligned incentives are rarely sufficient on their own. To remedy this problem, we create our own alignment in incentives by keeping track of who said what, and by lowering our trust in those who provided unhelpful information. In turn, this monitoring motivates speakers to be diligent in providing us information, creating a social alignment of incentives.

Because we are able to track others' commitments, and adjust our trust accordingly, most of human communication isn't cheap talk but costly signals, as we pay a cost if our messages turn out to have been unreliable. Arguably, it is this dynamic of commitment that has allowed human communication to reach its unprecedented scope and power. However, this ability to keep track of who said what, as well as to figure out whether a speaker's incentives are more or less aligned with ours, depends on having access to a wealth of information. Through the majority of our evolution, we would have known for most of our lives most of the people we interacted with. As a result, we would have had plenty of information to recognize aligned or misaligned incentives; to spot overconfident, unreliable, deceitful individuals; and to adjust accordingly how we value their commitment.

Ironically, even though we are now inundated with more information than ever, we know very little about many of the people whose actions affect us most. What do we know of the people who make sure the products we buy are safe, who operate on us, who fly our planes? We barely know the politicians who govern us, besides what we can glean from scripted speeches and carefully curated insights into their personal lives. How are we supposed, then, to decide who to trust?

One of the organizing principles of open vigilance mechanisms is that, in the absence of positive cues, we reject communicated information: by default, people are conservative rather than gullible. The same applies to trust. If we don't know anything about someone—indeed, if we don't even know who they are—we won't trust them. A first step to establish trust is thus to be recognized, as an individual or an entity. This is why name recognition matters in politics, and branding matters in marketing.[40]

Obviously, a name isn't enough. To convince others of the reliability of our messages, we have to do more. As we've seen in the preceding chapters, good arguments, having access to relevant information, or having performed well in the past can make a speaker more convincing. However, trustworthiness is often a bottleneck. Many arguments fall flat if we don't accept their premises on trust (e.g., that such and such study showing no link between autism and vaccines exists and is reliable). Even the best-informed, most competent speaker shouldn't be listened to if we don't think they have our interests at heart. Research on advertising and political behavior lends some support to the importance of trustworthiness. As we will see in chapter 9, celebrities help sell products mostly when they are perceived as expert in the relevant domain, but trustworthiness matters even more.[41] To the extent that the perceived personal traits of politicians weigh on voting decisions, a study suggests the most important trait is "how much the candidate really cares about people like you"—in other words, whether you think their incentives are aligned with yours.[42]

The importance of trustworthiness is also highlighted by the damage done when it breaks down. As a rule, the positive effect that association with a (competent, trustworthy) celebrity might have for a product is swamped by the fallout if negative information about the celebrity emerges.[43] For example, following the revelations about Tiger Woods's philandering, three brands that had recruited him—Pepsi, Electronic Arts, and Nike—lost nearly $6 billion in market value.[44] Similarly, politicians are barely rewarded for keeping their electoral promises (which they do most of the time, at least in democracies), but they suffer high electoral costs when they are convicted of corruption (as they should).[45]

The relative paucity of information available to help us decide who to trust means that, if anything, we do not trust enough—a theme I will return to in chapter 15.

7

WHAT TO FEEL?

IN THE EARLY 1960S, Tanganyika was in a state of flux. It had declared its independence from British rule in 1961 but remained part of the Commonwealth until 1962, when it shed that last link with the United Kingdom. A year later, Tanganyika would unite with neighboring Zanzibar to become the modern state of Tanzania.

But in 1961, politics was not the only area of turmoil in Tanganyika. In the Buboka district, on the western shore of Lake Victoria, children were behaving weirdly. It all started on January 30, when three teenage girls, pupils at the same boarding school, were beset with sudden, uncontrollable bouts of laughing and crying that lasted several hours.[1] A year later, nearly one hundred pupils were affected. The school was forced to close down. The pupils went back home, spreading further their unusual and upsetting behaviors. Over the coming months, the contagion would affect hundreds of young people throughout the district.

Outbreaks of bizarre behavior are hardly novel, and they are far from over.[2] In 2011, dozens of teenage girls from Le Roy, a small town in upstate New York, were affected for months on end with symptoms similar to those that had stricken the pupils in Tanganyika fifty years earlier.[3]

When describing these events, it is difficult to avoid analogies with epidemiology: outbreak, spread, affected, contagion. The two doctors who reported the events happening in Tanganyika described them as "an epidemic of laughter."[4]

The same analogies are often used to describe how people act in crowds. In the late nineteenth century, contagion became the dominant explanation for crowd behavior. Gustave Le Bon wrote that "in crowds, ideas, sentiments, emotions, beliefs possess a power of contagion as intense as that of microbes."[5] His colleague Gabriel Tarde noted that "urban crowds are those in which contagion is fastest, most intense, and most powerful."[6] The Italian Scipio Sighele suggested that "moral contagion is as certain as some physical diseases."[7]

It is hardly surprising that the analogy between the spread of emotions and pathogens flourished at that time. The second half of the nineteenth century was the golden age of the germ theory of disease, when it allowed John Snow to curb cholera epidemics; Louis Pasteur to develop inoculation against rabies; and Robert Koch to identify the agents that cause anthrax, cholera, and tuberculosis.[8]

The contagion analogy, as the germ theory of disease, would only grow more successful. It has been used to describe panics, such as the (supposed) reactions to Orson Welles's "War of the Worlds" broadcast, when thousands of listeners, thinking the Martians had landed, are supposed to have fled in panic: the "first viral-media event."[9] Armies are depicted as routing in "a contagion of bewilderment and fear and ignorance."[10] Nowadays, talk of contagion and viruses has become ubiquitous when describing the effects of social media—take the phrase *viral marketing*, for instance—and not only in the popular press. In 2014, two articles were published in the prestigious *Proceedings of the*

National Academy of Sciences that attempted to detect and manipulate "emotional contagion [in] social networks."[11]

There are indeed tantalizing parallels between the way diseases, on the one hand, and emotions or behaviors, on the other, spread. People do not voluntarily emit pathogens. Similarly, the expression of emotions or such bizarre behaviors as uncontrollable laughing or unstoppable dancing are not under voluntary control. People do not choose to be affected by pathogens. Similarly, people do not make a conscious decision to start laughing or crying when they see others do the same—indeed, in many cases they actively resist the urge. Some pathogens at least are extremely difficult to fight off. Likewise with emotions and behaviors in crowds: "Few can resist [their] contagion,"[12] suggested Nobel Prize–winning author Elias Canetti. Finally, pathogens can have dire consequences. Similarly, behavioral or emotional contagion can (or so it is said) "make of a man a hero or an assassin";[13] it can lead individuals to "sacrifice [their] personal interest for that of the collective."[14]

If nineteenth-century psychologists relied chiefly on crude observations of crowd behavior, their successors have performed impressive experiments showing how fast and automatic the reaction to emotional signals can be. Recording the movements of their participants' facial muscles, psychologists John Lanzetta and Basil Englis showed that seeing someone smile or frown immediately leads to the same muscles being activated in the observer's face.[15] Later, psychologist Ulf Dimberg and his colleagues showed that this automatic imitation happened even when the emotional expressions were presented so quickly that they could barely be consciously registered at all.[16] This almost instantaneous activation of facial muscles is believed to be a sign of contagion: people automatically take on the facial expressions of those they observe, which leads them to feel the same emotion. Psychologist Guillaume Dezecache and his colleagues have even shown this mimicry can extend to third parties: not only do observers take on the emotional expressions of those they observe, but those who observe the observers also adopt the expression.[17]

With such results, it is hardly surprising that one of the most influential books in the field of emotional communication, by psychologists Elaine Hatfield, John Cacioppo, and Richard Rapson, should be titled *Emotional Contagion*, and bear on the power of "rudimentary or primitive emotional contagion—that which is relatively automatic, unintentional, uncontrollable, and largely inaccessible to conversant awareness."[18]

PASSION WITHIN REASON?

From the evolutionary perspective I have adopted in this book, emotional contagion is implausible. If emotions were truly contagious, if they forced irrepressible mimicry, they would be too easily abused. Cheaters could laugh until those they have cheated laughed with them. Mortal enemies could get their opponents to empathize with and care for them. If our emotions were so easily manipulated, we would be much better off not paying any attention to emotional signals.

Contemporary emotion researchers, such as Hatfield and her colleagues, are quick to point out the limits of the basic view of contagion, in which like begets like, as scholar Alfred Espinas suggested in the nineteenth century: "The representation of an emotional state gives rise to this same state in those who witness it."[19] Such contagion makes no sense whatsoever for some emotions. Take anger. We express anger to impress on others that we have been wronged, and that this better not happen again.[20]

If the only effect of expressing anger were to make others angry in turn, expressing anger would be counterproductive.

We could thus imagine a version of contagion that would be much less strict than "like begets like." An emotion would simply have to provoke a reaction, even if that reaction were different from the initial emotion. But that wouldn't solve the problem raised by contagion. If displays of anger consistently led the onlookers to yield to the angry individual, the weakest among us could show their anger and make any opponent submit, regardless of their relative strengths.

There has to be something that keeps emotional signals broadly reliable, that is, beneficial on average for those who receive them. Charles Darwin, who devoted a book to the expression of emotions, was well aware of the problem. In the case of blushing he cites one of his colleagues, Thomas Burgess, who suggested that blushing serves the function of putting our shameful wrongdoings in full view. For Burgess, the creator provided the soul with the "sovereign power of displaying in the cheek, that part of the human body which is uncovered by all nations, the various internal emotions of the moral feelings whenever they are infringed upon either by accident or design."[21] Blushing would be an honest signal because of a direct line between the soul and our cheeks, one that could not be tampered with by conscious volition. Surprisingly enough, the common answer to why emotional signals remain reliable hasn't changed much since Burgess attributed this reliability to the benign hand of the lord.

Economist Robert Frank wrote the most thought-provoking book about the function of emotions since Darwin: *Passions within Reason*.[22] In this work, he argues that displaying emotions can be rational. Consider the problem raised by making credible threats of retaliation. To prevent others from wronging us, we want them to believe that if they wrong us, we will retaliate.

However, as I explained in the preceding chapter, once we have been wronged, retaliation is often not the most sensible option. Imagine you get scammed when buying a cheap product online, and the seller is adamant that they won't issue a refund. You could sue them, but that might cost a lot of money, and it is sure to take up a lot of time. It is often more sensible to just drop it. Knowing this, scammers can take advantage of you. But if you could convince scammers that you would retaliate, regardless of the costs, then you wouldn't get scammed—and so you wouldn't even have to retaliate.[23]

For Frank, emotions and their expression evolved by solving these types of problems. Anger would have evolved to express one's commitment to retaliate if wronged, regardless of the costs of retaliation. The obvious question, then, is what makes the expression of anger credible? For Frank—and others before and after him—the answer is that the expression of emotions, like our reaction to these expressions, is automatic, outside of conscious control: "If all the facial muscles were perfectly subject to conscious control, facial expressions would be robbed of their capacity to convey emotional information."[24] Other cues are also described as being intrinsically honest because they are outside of conscious control: pupil dilation as a cue to arousal, blushing as a cue to guilt.

The circuit of emotional contagion would thus be complete: people can afford to react automatically to emotional signals because emotional signals are sent automatically and thus are impossible to fake. By this means, emotions could spread until they affect a whole crowd, in an avalanche of automatic and compelling signals.

From an evolutionary point of view this reasoning does not hold water, as it is irrelevant whether or not a signal is sent consciously. Take Thomson's gazelles trying to communicate to wild dogs that they are too fit to be caught. Instead of stotting, which requires some energy, why not emit some kind of call? Whether or not the call was under the gazelles' conscious control would not affect its reliability: unfit gazelles would soon evolve to give the same call, voluntarily or not, and dogs would pay the call no heed.

Similarly, if some behaviors or emotional displays were able to reliably elicit a reaction in their audience, individuals would evolve to send these signals whenever it was in their interest to do so, even if that meant sending unreliable signals. The gazelles' stotting is a reliable signal because it is simply impossible for unfit gazelles to evolve the capacity to stot convincingly. By contrast, there is no such obstacle for emotional signals, however automatic they are. So why not display anger even when we would never retaliate? Why not blush even if we wouldn't hesitate for one minute to repeat our shameful actions?

If automaticity is no guarantee of reliability, why do we pay any attention to emotional signals? What keeps them honest?

Emotional Vigilance

The answer lies in clearing up the confusion between two closely related concepts: automatic and mandatory.[25] A cognitive mechanism is automatic if it functions outside of conscious control. Fortunately, this is true of the vast majority of cognition, most of the time: we couldn't consciously focus on all the steps necessary to make sense of an utterance or interpret a visual scene. A cognitive mechanism is mandatory if it cannot help but run its course once the right stimulus is present. Mandatory mechanisms would be like reflexes: if the doctor's hammer hits you on the right spot under the knee, you will always raise your foot.

It is tempting to think that if a cognitive mechanism is automatic, then it has to also be mandatory. That's only because we focus too much on conscious control; in fact, if most cognitive mechanisms are automatic, very few, if any, are mandatory.[26] Seeing a slice of scrumptious chocolate cake makes most people hunger for it. This reaction is hard to repress—it is automatic—even when we're on a diet (especially when we're on a diet). However, the same slice of chocolate cake might elicit only disgust after a heavy meal capped by two portions of cheesecake. Again, this reaction would be wholly automatic. Yet, because the same stimulus can yield opposite reactions in different contexts, neither reaction is mandatory.

If our reactions to emotional signals aren't mandatory, then there is room for what Guillaume Dezecache, Thom Scott-Phillips, and I have called *emotional vigilance*—mechanisms of open vigilance dedicated to emotional signals.[27] Even if they do so unconsciously, people should be able to adjust their reactions to emotional signals so as to stop responses that are not in their best interest. The application of this emotional vigilance would then provide incentives for senders to avoid sending unreliable emotional signals.

How should emotional vigilance function? There is likely not a one-size-fits-all recipe. Emotional vigilance should be attuned to the properties of different emotions. For instance, disgust might offer fewer opportunities for manipulation than anger: think of how useful many (all?) would find it to make anyone they want submit to them. By contrast, making people disgusted appears less useful—except maybe to get all of that scrumptious chocolate cake for yourself. Still, when reacting to emotional signals, the following three factors should be relevant across all emotions: what our prior beliefs and plans are, in what context the signals are produced, and whether the sender is trustworthy.

Even babies and toddlers—not the creatures one would grant with the most sophisticated emotional control—can take these factors into account when reacting to emotional signals. Not many parents will be surprised to hear that children are masters of selective ignorance. They pay attention to what their parents say only when it suits them. An experiment conducted by psychologist Catherine Tamis-LeMonda and her colleagues offers a nice demonstration with eighteen-month-olds.[28] The babies had to choose whether or not to walk down a slope. Their mothers were either encouraging them to walk or telling them not to, with a mix of emotional signals. The babies could not avoid the signals because their mothers were right in front of them, gesticulating and making faces. Indeed, they understood their mothers perfectly. For slopes that were neither too steep nor too flat, the babies paid attention to the mothers: only a quarter ventured down the slope when the mother was sending negative signals, while three-quarters went down if she was sending positive signals. Otherwise, the babies completely ignored their mothers. If the slope was perfectly safe—only a few degrees of inclination—the babies went for it even if their mother was urging them not to. If the slope was clearly dangerous—a fifty-degree angle—the babies stopped, irrespective of what the mother was signaling (an experimenter was there to catch them had they decided otherwise; no babies were hurt in the making of this experiment). The babies were engaging in a simple form of plausibility checking.

Young toddlers also understand when an emotional display is justified. In their article "Cry Babies and Pollyannas," psychologists Sabrina Chiarella and Diane Poulin-Dubois describe an experiment, also conducted with eighteen-month-olds.[29] The babies were shown videos of an actress expressing either a justified emotion (happiness after being handed a nice toy) or an unjustified one (sadness in the same conditions). The babies were more curious about the unjustified display of sadness, looking back and forth between the object and the actor, trying to figure out what was happening. They also expressed less concern, and were less likely to call for help, when the emotion seemed unjustified.

A similar experiment with toddlers—three-year-olds—revealed that they not only adjusted their reactions as a function of whether or not the emotional displays were justified but also held senders of unreliable signals accountable. Psychologist Robert Hepach and his colleagues had the toddlers interact with adults who expressed consistently justified or unjustified emotions.[30] The adult would express distress if her sleeve (rather than her hand) got caught under the heavy lid of a box, she would cry after a drawing of hers was slightly dented (rather than torn apart), and she would complain at having received a fair (rather than unfair) share of marbles in a game. Later on, the adult would start crying from behind a screen. When that adult had consistently sent unjustified signals, only a third of the toddlers checked on her, whereas more than 80 percent did so when her complaints had been well founded. In a later task, the toddlers were less likely to help the adult who had sent unreliable emotional signals.

Here lies the key to the stability of emotional signals: many are rejected outright, and those who abuse the signals end up paying the price. Abusers might not be punished in the typical sense of being physically punished, but their reputation suffers, as does the reputation of those who break more explicit commitments. Senders of unreliable emotional signals are trusted less when they send emotional signals, and possibly when they employ other forms of communication as well.

What about adults, then? Don't the experiments reported earlier in the chapter show that adults inevitably mimic the emotions they perceive in others? Could humans lose the ability to discriminate between reliable and unreliable emotional signals as they grow up?

No. Adults also adjust their reactions to emotional signals as a function of their source, and of the context in which the signals are emitted. Lanzetta and Englis had shown that participants automatically mimic the smile or frown expressed by a confederate, but only when the participants expected to cooperate with the confederate later on. When the participants expected to compete with him instead, they tended to show opposite reactions, smiling when the confederate received a shock and frowning when he was rewarded—what Lanzetta and Englis called *counterempathy*.[31]

Source effects have been reported in many experiments. Tears are taken to indicate sadness more reliably if they are shed by an adult rather than a toddler.[32] Women do not mimic the expressions of those who behave unfairly toward them.[33] Men express positive emotions when others show fear, and negative emotions when others show joy—if the others are fans of a rival sports team.[34] Even catching yawns, a seemingly perfect example of irresistible contagion, is not as reflexive as it seems: people are more likely to start yawning when they see people they know, rather than strangers, yawn.[35] And, like toddlers, adults increasingly mistrust those who mispresent their emotions—for instance, people who feign anger to obtain strategic advantages in negotiations.[36]

CONTAGION IS A CATCHY BUT MISLEADING ANALOGY

Our reactions to emotional signals might be automatic—we do not consciously control our emotional reactions—but they are far from being mandatory. Instead, they adapt according to a number of factors, including our preexisting plans or beliefs, the context, and the source's credibility. This calls into question the contagion analogy.[37] We evolved neither to send nor to receive pathogens—indeed, a good chunk of our evolution is devoted to avoiding the effects of pathogens. By contrast, we did evolve to send and receive emotional signals.[38] It therefore makes little sense to talk about contagion when talking about people's reactions to emotions.

Describing the transmission of emotions as resulting from contagion is merely giving a new name to the phenomenon of transmission, without any explanatory purchase: no facet of the phenomenon is better understood thanks to the analogy (on the contrary!).[39] Because contagion from actual pathogens is relatively well understood, this sleight of hand provides an illusion of understanding, but in fact pathogen contagion and emotional communication have more differences than commonalities.[40]

What about the costs, though? Can't contagion, whether by pathogens or by emotional signals, be costly for those affected? How can we reconcile the idea that our reactions to emotional signals are adaptive, fine-tuned to protect us from unreliable senders, with the epidemics of weird behaviors, or the way crowds are supposed to turn individuals into bloodthirsty hooligans or panicked sheep? In fact, the current adaptive perspective is not incompatible with the supposed cases of emotional contagion. Indeed, this perspective helps us understand why emotional expressions sometimes powerfully affect people, and sometimes have no effect whatsoever.

In the various cases of "mass psychogenic illness," such as laughter epidemics, what clearly sets apart the contagion perspective from the current adaptive view are the predictions regarding who the behaviors spread to. Pathogens spread to whoever is most in contact with the infectious agent. If the contagion analogy were accurate, we should expect a similar pattern with emotional signals. By contrast, the adaptive perspective predicts that influence should be heavily constrained by the state of mind of those who perceived the signals, and by their relationship with those who send them. Supporting the emotional vigilance perspective, the abnormal behaviors that characterize mass psychogenic illness rarely, if ever, spread outside a small coterie of people who know and trust each other. The symptoms typically affect a few dozen people at most, all of them members of the same group—pupils in a school, workers in a factory, inhabitants of a small village.[41] In contemporary cases of mass psychogenic illness, the area soon swarms with journalists, government representatives, experts, and rubberneckers. None of them is ever affected. In most cases, only one gender or age-group is affected. In high schools, the weird behaviors are communicated along the typical fault lines of teenage social life: cool kids who hang out together are affected first, followed later by less popular teens.[42]

Whether or not an individual starts displaying bizarre behaviors is a function of their existing relationship with those who already exhibit the symptoms, but also of their prior mental states. Behaviors that are truly harmful—violence toward others, serious self-harm—do not create mass psychogenic illnesses. Instead, we observe behaviors such as dizziness, jerking movements, or laughing. Moreover, people affected by mass psychogenic illnesses might derive some benefits. Those who experience these symptoms are likely to have suffered unusual stress, and the symptoms might allow them either to get out of a bad situation or at least to attract attention to it. In Tanganyika, the outbreaks mostly affected children caught between their traditional culture and that imposed by the nuns running their boarding schools. Factories affected by mass psychogenic illnesses tend to have particularly poor working conditions—and more people start exhibiting the symptoms when the possibility of litigation and compensation emerges.[43] For example, in the early 1980s, after a bizarre (but genuine) epidemic had taken over Spain, the government started offering compensation to those affected. Among the patients who exhibited not physical but psychiatric symptoms, the compensations, psychiatrists noted, "introduced a certain mimicry into the symptomatology," with some people imitating (probably unconsciously) the symptoms of those deemed by their doctors eligible for compensation.[44] By contrast with the weird behaviors that characterize mass psychogenic illnesses, the spread of pathogens is quite insensitive to the amount of harm caused: finding it really inconvenient to get the flu hardly prevents us from catching it.

The pattern of mass psychogenic illnesses—who tends to be affected by them, after contact with whom—is thus much better explained by taking an adaptive stance, in which our reactions to others' emotional displays are filtered by emotional vigilance, than by the contagion analogy. What about crowds, then? The answer here is simple. The view of crowds as passive herds subject to currents of violent passions is simply wrong, with no basis in fact.[45]

RATIONAL CROWDS

The conventional reactionary narrative describes the French Revolution as a "dictatorship of a mob," whose "proceedings, conforming to its nature, consist in acts of violence, wherever it finds resistance, it strikes."[46] Historian George Rudé restored the truth in *The Crowd in the French Revolution*.[47] If more than a hundred people die when the Bastille is taken, nearly all of them are revolutionaries. Rudé even wonders why "the angry and triumphant crowds" had been so restrained and had not killed more than a handful of guards. Two months later, the mob takes over Paris's city hall. Citizens tear apart official documents but leave a huge pile of money sitting there untouched. In July 1791, a crowd fifty thousand strong marches on the Champ de Mars. They do so mostly peacefully, while the National Guard, called in to control them, kills dozens of protesters. Throughout the revolutionary years, mobs of women take over warehouses or shops selling sugar. Instead of looting them, the women ask for a discount. Most of the defenseless victims of the revolutionary crowds are the prisoners killed during the September Massacres, but this slaughter is neither completely irrational nor wholly indiscriminate. Paris is attacked from all sides by foreign powers; most able men, and all the weapons, are at the front, leaving the city eminently vulnerable to an attack from within. And many prisoners are spared, such as those whose only crime is to be in debt, as well as women.

The strikers of the late nineteenth century, protesting their low wages and dangerous working conditions, scaring Le Bon, Tarde, and other crowd psychologists, were largely harmless. Out of twenty-seven hundred strikes, fewer than a hundred turned violent, and the crowds killed a grand total of one person (a nasty supervisor already hated by the workers).[48] The strikers were far more likely to be killed by guards and policemen than they were to kill anyone. The crowds' tameness even led anarchists to complain about the strikers' stupidity, citing crowd psychologists in support of their cause. Whether they are too violent or too tame, crowds are (wrongly) perceived as gullible.[49]

These broadly rational and surprisingly restrained actions are hardly a specialty of French crowds.

The peasants who revolted in fourteenth-century England took over manors, castles, and churches. Rather than wanton looting and killing, they were mostly content with burning the documents that held them in debt or bondage.[50]

Historian Aoki Koji documented more than seven thousand cases of popular protest in Tokugawa Japan (the era extending from 1600 to 1868). Only 2 percent of these uprisings led to deaths among those targeted by the protests.[51]

In 1786, thousands led by Daniel Shays took arms in the state of Massachusetts against the economic and political order. This insurrection, and others like it, terrified the framers of the U.S. Constitution.[52] Yet these revolting crowds proved largely toothless: Shays' Rebellion did not cause a single casualty, and most rebels ended up signing confessions to be granted amnesty.

In 1966, "spontaneous mobs" of Red Guards formed in the Chinese city of Wuhan.[53] These crowds targeted the houses of twenty-one thousand "monsters and ghosts"—people who were believed to oppose the Cultural Revolution. No one dared oppose the mobs, which could have killed at will; yet the Guards let 99.9 percent of their targets live.

That people in crowds can do horrible things, from lynching to sexual assault, is undeniable. The point here isn't to pass moral judgment on crowds and the individuals who constitute them, but to understand their dynamic. If crowds were truly animated by "contagious transports, irresistible currents of passion, epidemics of credulity," they should be much more consistently violent, unable to show restraint, wholly irrational.[54] Instead, crowds often eschew violence altogether or, failing this, consistently show discrimination in their actions, attacking specific targets and sparing others, using controlled strategies rather than all-out rampage.[55] Even morally depraved attacks need not be irrational: some people are ready to seize any chance to steal and assault (often specific) others. Their actions are not driven by "irresistible currents of passions" but by the opportunity to act with relative impunity that crowds provide.[56]

The same picture emerges when we look at cases of supposed panic.[57] Some instances of panic are simply invented—subsequent analyses showed that very few people panicked when they heard the infamous "War of the Worlds" radio program.[58] Even real, terrifying events, such as natural disasters or air raids, do not cause widespread panic.[59] Likewise, during warfare, panic leading to "serious disorganization . . . of combat units . . . was found to be exceedingly rare."[60] Instead of contagious panic we find in threatened crowds the same heterogeneity as in any other crowd. Human factors researcher Guylène Proulx and her colleagues analyzed the accounts of survivors from the 9/11 terrorist attacks on New York, people who were in the World Trade Center towers when they were hit by jumbo jets.[61] Mass panic, with everybody rushing for the exit, would be an understandable reaction in these circumstances. Instead, less than a third of firsthand accounts described others as being "momentarily panicked." The majority of survivors thought others were calm, and a substantial minority were helpful. Guillaume Dezecache and his colleagues observed a similar pattern in the reaction of the victims of the attacks on the Bataclan in Paris.[62] Even as terrorists were targeting them with automatic weapons, those trapped in the theater performed more pro-social actions (comforting others, for instance) than antisocial actions. Moreover, the antisocial actions—such as pushing others to get to the exit—were driven by rational (albeit selfish) factors, such as the possibility of escape, rather than sheer panic.

The reactions of the survivors of the 9/11 and Bataclan attacks are no exception. In every emergency situation, a few people do react as if panicked—rushing for the exit, pushing anybody who is in their way. But panic is a misleading description, as it suggests that the behavior is irrational and easily transmitted. Instead, fleeing by any means available in the face of shots, fire, or other perceived threats, selfish as it might be, is hardly irrational. And the panic doesn't spread: most people behave calmly enough, and many help others, in particular the most vulnerable.[63]

The popular image of crowds, whether they are rioting or panicking, suffers from the "illusion of unanimity," an intuition that all crowd members are going to behave in the same way.[64] Even when crowd members share an ideology, their behavior is heterogeneous: they do not necessarily follow each other's actions or the leader's demands.[65] If there are no "irresistible currents of passions," crowd members do influence each other, but mostly within small groups: people who have joined the crowd together, who know each other well, and who can more easily trust each other's reactions.[66] The only exception to this pattern is that of retreating soldiers. The less the soldiers know and trust each other, the more likely they are to emulate those who start fleeing, leading to a full-scale rout. But this has nothing to do with emotional contagion. When soldiers infer, rightly enough, that they cannot trust others to hold the line, they try not to find themselves in that least enviable of positions: being the last one to flee.[67]

Instead of indiscriminately catching whatever emotion we happen to witness, we exert emotional vigilance—even when we are in the middle of a crowd. For us to react to emotional signals in the way intended by their sender, the reaction has to suit our current plans and mental states, and the sender has to be someone we like, who has not proven unreliable in the past, and whose emotion seems justified. Otherwise, we might not react at all, or we might react in a way opposite to that intended—rejoicing in someone's pain, or being angered by a display of anger.

8

DEMAGOGUES, PROPHETS, AND PREACHERS

EVOLUTION MAKES GULLIBILITY MALADAPTIVE. So as not to be abused by senders of unreliable messages, we are endowed with a suite of cognitive mechanisms that help us decide how much weight to put on what we hear or read. To do so, these open vigilance mechanisms consider a number of cues: Are good arguments being offered? Is the source competent? Does the source have my interests at heart?

When it comes to large audiences, for better or worse these cues do not scale up well. Argumentation is most efficient in the context of a small group discussion, with its back-and-forth of arguments and counterarguments. When aiming a speech at millions, speakers have to resort to common denominators, and they cannot anticipate the many objections that are certain to arise. Demonstrating competence to a wide audience is difficult: with limited knowledge and attention span, how are listeners supposed to know who is the most competent politician or economist? Similarly, credibly displaying one's goodwill is easier said than done, as building trust is best done slowly, one individual at a time.

When our more sophisticated open vigilance mechanisms do not operate, we are left with plausibility checking. Plausibility checking is always on, ever vigilant. As a result, it should exert a disproportionate influence on mass persuasion, making it tremendously difficult to change people's minds. At best, mass persuaders can hope to spread messages that conform with the public's preexisting plans and beliefs. With a bit of work, they will be able to affect their audience at the margin, on issues for which the audience is ambivalent or had weak opinions to start with.

Yet many have granted prophets the power to convert whole crowds, propagandists the ability to subvert entire nations, campaigners the skill to direct electoral outcomes, and advertisers the capacity to turn us all into mindless consumerists. Could this be all wrong?

DEMAGOGUES

If ancient Athens is the blueprint for democracy, Cleon is a blueprint for democracy's "worst enemies": the demagogues.[1] As described by politician Michael Signer, Cleon "took over the Athenian government, came within a hair of executing a powerful playwright who dared challenge him, attempted the mass murder of the inhabitants of a vanquished island, launched reckless military expeditions, and brought Athens into a war that ultimately would defeat its democracy for a time."[2] According to his critics, Cleon had become "very powerful with the multitude" thanks to his charisma, in particular his powerful voice, with which he harangued the Athenian demos.[3] The power of demagogues such as Cleon is often taken to be the foremost illustration of the gullibility of the masses.

With hindsight, there is no doubt that some of Cleon's choices were morally repugnant or strategically dubious. But the real question here is: Was Cleon able to use his charisma to talk the Athenians into making decisions that were good for him but bad for them?

The most infamous of the decisions attributed to Cleon is the order to wipe out the inhabitants of Mytelene in reprisal for revolting against Athens, the archetypal example of a bloodthirsty demagogue driving the populace to evil deeds. But did Cleon really have to use his charismatic powers to persuade the Athenians to commit such an atrocity? It seems unlikely. Mytelene had betrayed Athens by conspiring with its enemy, Sparta.[4] It had recruited other cities to the rebellion. Given the standards of the time, a brutal punishment for such deeds would be expected. Ironically, the events better illustrate the weakness of the demagogue's hold. The day after a trireme had been sent to execute the order, the debate was reopened, and Cleon's opponent, Diodotus, persuaded his fellow Athenians that, for practical reasons, the population should be spared.[5] Another trireme was sent to intercept the first one, which it did successfully. The oligarchs were removed and the rest of the population spared.

Not only was Cleon's charisma too weak to counteract sound arguments, it also was unable to protect him from ridicule. When Aristophanes pilloried Cleon in his plays, the crowd was amused, not angered—the same crowd that Cleon, as a good demagogue, was supposed to have "entirely at his disposal."[6] Indeed, when Cleon raised trumped-up charges against Aristophanes, a popular jury sided with the playwright.

If the people's support for Cleon was far from unconditional, it still was, by and large, genuine: after all, the Athenians made Cleon a general and voted for many of his policies. But his power was not unearned. Cleon's economic policies seem to have benefited the poor majority.[7] Cleon's influence was not due to extraordinary feats of persuasion but rather to the fact that he possessed the "true demagogue's tact of catching the feeling of the people."[8] Not being an aristocrat, he was free to enact populist policies, "challeng[ing] the authority of wealth and unexamined tradition."[9] By and large, Cleon's powerful voice reflected, rather than guided, the people's will—for better or worse.

Other demagogues—such as the long line of American populists, from William Jennings Bryan to Huey Long—have relied on the same strategy, gaining political power not by manipulating crowds but by championing opinions that were already popular but not well represented by political leaders. Even the most infamous of demagogues, Adolf Hitler, fits this pattern.

Thanks to a wide variety of sources—from diaries to the reports of the Nazi intelligence services—historian Ian Kershaw has gained an intimate knowledge of German public opinion under the Nazis.[10] In *The Hitler Myth*, he describes how Hitler was perceived by ordinary Germans throughout his political career, and how he gained, for a time, broad popular support.[11] For Kershaw, the key to Hitler's electoral success in 1933 was that he "embodied an already well-established, extensive, ideological consensus."[12] In particular, Hitler surfed on a wave of virulent anti-Marxism, a cause he shared with the church and the business elites.[13]

From 1927 to 1933, Hitler used innovative campaign strategies, techniques that have now become commonplace. He flew across Germany so that he could reach more people. He used loudspeakers amplifying his voice to make the best of a full rhetorical arsenal. He gave hundreds of speeches to crowds large and small. Were these efforts successful? A careful study suggests they weren't. Political scientists Peter Selb and Simon Munzert found that Hitler's countless speeches "had a negligible impact on the Nazis' electoral fortunes."[14]

Once he had risen to power, Hitler's appeal waxed and waned with economic and military vicissitudes. He gained in popularity among those who benefited from his policies, and with the general public when painless military victories came in quick succession.[15] As early as 1939, however, as Germans tightened their belts for the war effort, discontent began to grow.[16] After the Nazi disaster that was the Battle of Stalingrad, support for Hitler disintegrated. People stopped seeing him as an inspirational leader, and vicious rumors started to circulate.[17] Even though it was a capital crime, from 1943 until his suicide in April 1945, many Germans were openly critical of Hitler.[18]

Far from shaping German public opinion, Hitler responded to it; as Kershaw put it, "More than any other exponent of propaganda, Hitler had an extremely sensitive awareness of the tolerance level of the mass of the population."[19] In order to gain control he had to preach messages that ran against his worldview. During his rise to power, Hitler downplayed his own anti-Semitism, barely mentioning it in public speeches, refusing to sign the appeal for a boycott of Jewish shops.[20] Like other demagogues, Hitler was unable to rely on his own powers of persuasion to influence the masses, but rather played on people's existing opinions.[21] As we will see later, the Nazi propaganda machine as a whole was barely more effective.

PROPHETS

The power of demagogues to influence the masses has been widely exaggerated. What about religious figures such as prophets? History suggests prophets are able to whip up crowds into the kind of fervor that leads to suicidal acts, from self-sacrifices to doomed crusades. Yet if one steps back for a moment it soon becomes clear that what matters is the audience's state of mind and material conditions, not the prophet's powers of persuasion. Once people are ready for extreme actions, some prophet will rise and provide the spark that lights the fire.[22]

In the mid-1850s, Nongqawuse became a powerful seer among the Xhosa, a pastoral people of South Africa.[23] She made grandiose prophecies: if the Xhosa obeyed her, "nobody would ever lead a troubled life. People would get whatever they wanted. Everything would be available in abundance. . . . All the people who have not arms and legs will have them restored, the blind people will also see, and the old people would become young."[24] Nongqawuse also told of a powerful army that would rise from the dead to fight off the British invaders. But to make their dreams come true, the Xhosa had to kill all their cattle and burn all their crops. Many did so, killing every single one of their cattle and burning their crops to the roots. Yet only death and famine came in abundance.

Isn't that a dreadful example of extreme gullibility and mass persuasion? The Xhosa had no good reason to trust Nongqawuse, whom nobody knew. She offered no sensible justification for the actions she urged, and the actions themselves seemed very costly. To the British observers, Nongqawuse had simply "play[ed] upon the credulity" of her people.[25] This account, however, omits crucial factors that make sense of the Xhosa's behavior.

The years 1856–1857 had seen an epidemic of "lungsickness" wipe out whole herds of cattle.[26] In these circumstances, killing the animals and eating them before they got sick starts looking like a reasonable option.[27] The importance of lungsickness in driving the cattle killing can barely be exaggerated: in areas not affected by the disease, not a single animal was sacrificed.[28] As historian Jeff Peires, whose research I rely on here, concluded, "Lungsickness was thus a necessary cause of the Xhosa Cattle-Killing."[29] To some extent, the same reasoning applies to the crops that an unusually wet season had rendered susceptible to blight.

DEMAGOGUES, PROPHETS, PREACHERS

Even in areas affected by lungsickness, people didn't blindly obey Nongqawuse. They started by killing one or two head of cattle, honoring a long sacrificial tradition.[30] They kept the most important animals to be sacrificed last.[31] When Nongqawuse's prophecies failed to materialize, people quickly grew disillusioned.[32] In some cases, what drove people to kill their cattle were the threats of chiefs, neighbors, or even relatives, who had lost everything and were looking askance at those who refused to make their own sacrifice for the supposed common good.[33]

Peires argues that "the Cattle-Killing was a logical and rational response, perhaps even an inevitable response, by a nation driven to desperation by pressures that people today can barely imagine."[34] Even if this conclusion might be somewhat exaggerated, Peires's research shows that Nongqawuse did not hold any magical sway over the Xhosa. Instead, the Xhosa who followed her lead were driven by necessity to extreme actions.

The Cattle-Killing movement was also one of contestation, of near revolt.[35] Until then, the Xhosa had tolerated their chiefs owning most of the cattle, as the chiefs could be relied on to share when times got tough. But the situation changed when aristocrats started selling their surplus cattle to British settlers instead of sharing them in communal feasts.[36] This motivated many commoners to push for cattle killing: not only were the cattle not theirs, but they couldn't even serve as "drought insurance" anymore.[37] By contrast, those who benefited from the cattle trade overwhelmingly opposed the killing.[38]

In this way at least, the Xhosa Cattle-Killing episode is typical of other millenarian movements. Over the centuries, a great many people have been touched by millenarianism, believing the end of the world to be nigh, and a much better world to be within reach. Like the Xhosa, those who shared those beliefs often engaged in seemingly senseless acts, such as the impoverished

Christians in Europe who followed the injunctions of prophets, took up the cross, and attempted to take back Jerusalem. Yet their actions were not the result of mass persuasion, often being led by more down-to-earth considerations.

By and large, poor people's millenarian movements in the European Middle Ages were driven by desperation and the hope for material gains. When the most successful of the poor people's crusades reached Jerusalem, their leader cried out to them, "Where are the poor folk who want property!"[39] For historian Eugen Weber,

> The trouble was that most of these brow-beaten folk were less interested in the millennium per se than in the extermination that would precede it: the overthrow of oppressors, the annihilation of clergy and Jews, the end of the rich and fat. Their ecstasies and eruptions brought not peace but a pickax. From the twelfth century to the sixteenth and the seventeenth, while eschatological excitement ran high, crusades turned into massacres, and spiritual aspirations turned into social and political insurrections.[40]

Other historians concur: millenarianism, this "religion of the oppressed," mostly arises "under conditions . . . of felt or experienced crisis—of oppression by a more powerful group, of extreme economic hardship, of fundamental social changes that leave particular social strata feeling threatened."[41]

Challenging the social order is the quintessential breaking of norms; as such, it requires strong justifications. Millenarian beliefs provide such justifications: it is fine to wreck things, as the world is going to end anyway, and something much better will follow. This is why millenarian beliefs can be found in so many movements of contestation, across different cultures. If the best-known millenarian beliefs are Christian, the idea long predates the New Testament—it can be found in Jewish or Zoroastrian texts—and it was developed largely independently in other religions such as Buddhism.[42] Moreover, Christian millenarianism has been adapted in a variety of ways by different populations, from the Xhosa in South Africa to the Taiping rebels in China, often in spite of rather than thanks to the efforts of missionaries.[43]

Millenarian prophecies are successful whenever and wherever they are convenient. They surface across vastly different cultures, when people mount radical contestations of the existing order. Even secular upheavals have their own versions of millenarianism, with invocations of a golden age the revolution will bring back after a period of chaos.[44] The market for prophecies of doom is driven by the demands of discontented crowds rather than by the supply of sly prophets.

PREACHERS

Prophets may not carry that much influence over the masses, but what about the religious figures that (mostly) don't rely on threats of imminent apocalypse? The cultural success of Buddhism (520 million followers), Christianity (2,420 million followers), or Islam (1,800 million followers) suggests that some preachers have been able to convert vast flocks to their creeds. And these triumphs are not restricted to centuries-old religions: the rise of Mormonism in the nineteenth century and the success of the New Religious Movements—from Krishnas to Moonies—in the twentieth show that similar, even if so far smaller-scale, feats can be repeated with modern audiences.

When considering how one individual's vision might be communicated to millions, or even billions, of followers, it is hard not to think that mass conversion must be at play. In the Bible, one of Peter's sermons is described as having "added that day about three thousand souls."[45] In the fourth century, historian Eusebius wrote, "At the first hearing whole multitudes in a body eagerly embraced in their souls piety towards the Creator of the universe."[46] Many twentieth-century historians share the view that exponential religious growth must require "successes *en masse*."[47] Similarly, the development of New Religious Movements has worried many observers, who accuse their leaders of brainwashing new recruits.[48]

These visions of mass conversion stem from a misunderstanding of compound interests: a small but regular growth yields huge amounts over long time periods. If you had invested $1 in the year 0, to get $2,420 million now ($1 for each Christian on earth), you would only need a constant yearly interest rate of a little over 1 percent. According to sociologist Rodney Stark, who compiled estimates by several historians, the number of Christians went from around a thousand in 40 CE to 34 million in 350 CE. Even though this was Christianity's period of most rapid expansion, the increase only translates into a constant annual growth rate of 3.5 percent.[49] To explain the spectacular rise of Christianity, from a handful of followers to dozens of million in three centuries, you only need each Christian to make a couple of new converts in their lifetime—not exactly mass conversion.

More recent religious movements have generated similar conversion rates. Stark's studies of the early Mormon Church yielded growth rates below 5 percent a year.[50] Sociologist Eileen Barker conducted detailed observations of how new recruits—often called Moonies, after the founder, Sun Myung Moon—joined the Unification Church.[51] Even though the Unification Church was one of the most popular of the New Religious Movements, its success rate was very low. Among people interested enough to visit one of the church's centers, "not one in 200 remained in the movement two years later."[52] Even among those who went on two-day retreats, "only 5 percent remained full-time members one year later."[53]

Far from preachers managing feats of mass persuasion, religious conversion is, with few exceptions, driven by strong preexisting relationships. Friends recruit friends, families bring other family members into the fold. The beginnings of the Unification Church in the United States, which have been studied in detail by Stark and his colleague John Lofland, follow this pattern. The movement was led by Young Oon Kim, who after years of trying her best to "win converts through lectures and press releases"[54] had only managed to recruit a dozen people, good friends of hers and their families. Since this pioneering work, the importance—indeed, the quasi necessity—of close personal ties for conversion has been repeatedly observed, from Mormons to Nichiren Shoshu Buddhists, or medieval Cathars.[55]

Even if people are recruited by friends or family, conversion can entail some social costs inflicted by those not already converted, ranging from misunderstanding to persecution. In these conditions, doesn't conversion reflect a feat of persuasion, getting someone to accept, on trust alone, a new set of beliefs, often accompanied by costly personal obligations? On the contrary, it seems that people who convert find something to their liking in their new group. Summarizing the literature on New Religious Movements, psychologist Dick Anthony notes that "the psychological and emotional condition of most converts improves rather than declines after joining."[56] Even costly behaviors can be beneficial. Mormons have to donate 10 percent of their income and 10 percent of their time to the church. Yet it is not too hard to see why some people would prefer to live in a community in which everyone shares so much, enabling Mormons to "lavish social services upon one another."[57] Even early Christians, who, at times, were at great risk of persecution, likely benefited from the support networks created by their adhesion to this new cult.[58] By contrast with these practical aspects, the apparently exotic beliefs associated with new religions play a minor, post hoc role. As economist Laurence Iannaccone put it, "Belief typically *follows* involvement. Strong attachments draw people into religious groups, but strong beliefs develop more slowly or never develop at all."[59]

New religious movements can grow by offering people a mode of social interaction they enjoy, without involving mass conversion. But what happens when a religion becomes ubiquitous or dominant? Aren't its priests, then, able to dictate the people's thoughts and behavior?

Throughout the Middle Ages, the Catholic Church attempted to impose on the European peasantry behaviors that weren't in their obvious interest, from regular church attendance and confession all the way to the tithe, a 10 percent tax on whatever the peasants gathered each year. Moreover, the church also spread beliefs supporting the existing, iniquitous status quo. Kings had a divine right to rule. Priests taught that a view of the rich as merely lucky rather than deserving was "akin to covetousness," the root of all evil.[60]

This is what Marxist scholars have called the dominant ideology: a worldview created by the upper classes, justifying their position, that they impose on the rest of the population.[61] For Marx and Engels, "The class which has the means of material production at its disposal, has control at the same time over the means of mental production, so that thereby, generally speaking, the ideas of those who lack the means of mental production are subject to it."[62] Getting wide swaths of the population to accept an ideology in which their misery is deserved, making them resignedly accept their fate, would be the most impressive feat of mass persuasion ever accomplished.

In line with this vision, the Catholic Church is often described as ruling supreme over the European Middle Ages. A mixture of deference, ignorance, and fear of hell would have enabled the church to make a sheepish population obey its injunctions and accept its doctrine.[63] Humbert de Romans, a thirteenth-century Dominican who preached in poor areas of southern France, provides a very different perspective on how this was working out for the church. Humbert was probably pretty good at his job—he later rose through the ranks to become the head of the Dominican order—yet he despaired of what he saw on the ground.

The church, for all its power, was barely able to get the poor to conform to the bare minimum of its doctrinal requirements: to be baptized, to know the "Our Father," to take communion once a year.[64] Humbert complains of people going to church only to "pass the night gossiping to each other, not only about vain subjects, but about evil and indecent ones."[65] What about special occasions? The flock did enjoy saints' days, yet it wasn't the church that benefited, but "inn keepers and prostitutes."[66] Believers also went on pilgrimages, the occasion for "more sin, sometimes, than a participant committed in all the rest of the year put together."[67]

The poor not only turned religious ceremonies into opportunities for debauchery but actively resisted any costly behavior the church attempted to impose. Humbert deplores "the neglect of penance or fasting" and "the reluctance to pay tithes."[68] Historian Emmanuel Le Roy Ladurie notes how "the conflict over tithes . . . runs like a thread through peasant protests; it constitutes, from Catharism to Calvinism, a common denominator, more obvious than any dogmatic continuity, which is often absent."[69]

How do the people's crusades fit into this picture of disobedience? Didn't the church manage to persuade thousands of poor people to perform the ultimate sacrifice on its behalf? As mentioned earlier, the poor often saw such crusades as an opportunity for pillaging, rather than a spiritual calling. In any case, these crusades weren't a brainchild of the church hierarchy.[70] Indeed, the church sometimes actively fought them, and for good reason, as the "eschatologically inspired hordes of the poor," after they had looted any Jewish dwelling in sight, "soon turned on the clergy."[71] At the height of the first Shepherd's crusade, "the murder of a priest was regarded as particularly praiseworthy."[72] It seems the shepherds didn't get the memo about dominant ideology.

As well as refusing to comply with the costly behaviors the church demanded of them, the medieval masses also rejected the bulk of Catholic doctrinal orthodoxy. Historians of ideas have noted how, all the way to the Enlightenment, "deep-seated and persistent paganism frequently camouflaged with the most superficial veneer [of Christianity]."[73] In his meticulous study of a French village in the thirteenth century, Le Roy Ladurie notes a variety of decidedly not-so-Christian goings-on. Upon the death of a household head, other members keep his hair and nail clippings, "bearers of vital energy," through which the house absorbs some of the deceased's magical properties. A girl's first menstrual blood is saved to be used later as a love potion. Umbilical cords are preciously preserved, as they are thought to help win lawsuits (no refunds allowed).[74] It is not surprising that Humbert, our Dominican preacher, would reproach peasants for being "much prone to sortilege" and "so obstinate, nay even incorrigible, that they simply cannot be stopped, either by excommunications, or by any other kind of threat."[75]

Among the beliefs squarely rejected by the poor were those supposed to make them docilely accept their condition. Humbert might have taught that covetousness was a sin, but he still remarked how "the masses are accustomed to regard the rich of this world as the lucky ones," and are eager to complain about "the poor state of church government," which they (often, rightfully enough) blame on "bad bishops."[76] Throughout the Middle Ages, the Catholic Church's efforts to impose unappealing beliefs and costly behaviors on the population through mass persuasion are a litany of failures.

The pattern observed in Catholic medieval Europe recurs in economically dominated classes throughout the world.[77] Far from having imbibed the dominant ideology, everywhere people practice "the arts of resistance" with the "weapons of the weak," to borrow the titles of two influential books by sociologist James Scott.[78] Even the strongest of power asymmetries, that between masters and their slaves, cannot make the subordinates accept their plight, which slaves keep fighting by any means available, ranging from "foot dragging, dissimulation, false compliance," all the way to "arson" and "sabotage."[79]

The dominant ideology thesis has a point: the dominant classes weave narratives of the status quo as the best of all possible worlds, their superior position well deserved. Oftentimes, these narratives crowd communication channels, from manuscripts to airwaves. But this does not mean that people farther down the social ladder are buying any of it. On the contrary, these narratives are resisted everywhere, and alternative narratives created—including millenarian visions when an opportunity for revolution arises.

9

PROPAGANDISTS, CAMPAIGNERS, AND ADVERTISERS

PROPAGANDISTS

While he was in jail, working on *Mein Kampf*, Hitler thought a lot about propaganda. He described the masses as credulous: a "crowd of human children," "feminine in its character" (children and woman had long been associated with emotionality and gullibility). Accordingly, effective propaganda must rest on "stereotyped formulas ... persistently repeated until the very last individual has come to grasp the idea that has been put forward."[1]

Once he had been elected—mostly on an economic, anticommunist platform—and had consolidated his power, Hitler, with the help of Joseph Goebbels and the propaganda ministry, turned his theory into practice. Together, they would develop the most ignominious mass persuasion attempt in history, in particular as it aimed to build up German anti-Semitism, vilifying Jewish people in movies, radio programs, books, posters, educational materials, and so forth.

How effective was this "propaganda barrage"?[2] To obtain fine-grained data on German anti-Semitism, economists Nico Voigtländer and Hans-Joachim Voth looked at surveys from 1996 and 2006.³ They measured whether anti-Semitism is still higher among Germans who were exposed to the Nazi propaganda machine, in particular those born in the 1920s and 1930s. They found that indeed these cohorts expressed stronger anti-Semitic sentiments: Germans born in the 1920s and 1930s agreed with statements such as "Jews have too much influence in the world" between 5 and 10 percent more than those born at any other time.

Even if Nazi propaganda is responsible for only a small share of the anti-Semitism currently present in Germany, it certainly seems to have contributed to the problem. But did it do so through brute repetition, as Hitler thought? Voigtländer and Voth looked at the regional variations in the availability of propaganda—how many people owned radios, the number of cinemas where they could watch propaganda films, and so forth. If mere repetition were effective, areas with greater exposure to propaganda should see the sharpest rise in anti-Semitism. In fact, the sheer exposure to propaganda had no effect at all. Instead, it was the presence of preexisting anti-Semitism that explained the regional variation in the effectiveness of propaganda. Only the areas that were the most anti-Semitic before Hitler came to power proved receptive. For people in these areas, the anti-Semitic propaganda might have been used as a reliable cue that the government was on their side, and thus that they could freely express their prejudices.⁴ Another study that focused on the effects of radio broadcasts yielded even stronger results: radio propaganda was "effective in places where antisemitism was historically high," but it had "a negative effect in places with historically low antisemitism."⁵

Ian Kershaw, the historian we encountered in the last chapter scouring the records of Nazi Germany to understand Hitler's popularity, also analyzed the effectiveness of Nazi propaganda. He reached similar conclusions. The Germans didn't heed the calls to boycott Jewish stores and to ostracize the Jews more generally. It was only through "terror and legal discrimination" that the Nazis achieved "the economic (and increasingly social) exclusion of the Jews from German life."[6]

Kershaw argues that other dimensions of Nazi propaganda were even less effective than the attempts to turn all Germans into rabid anti-Semites. The push to enforce compulsory euthanasia of the handicapped was widely resisted.[7] Attacks on communism appealed to those already on the right but were an "almost unmitigated failure . . . among German industrial workers," communism's natural constituency.[8] Indeed, Nazi propaganda failed to persuade the majority of German workers to contribute willingly to the war effort, many electing to resist through absenteeism.[9] As soon as the war took a turn for the worst—after Stalingrad in particular—messages from the propaganda ministry fell on deaf ears. Goebbels's "unvarying message of victory was becoming monotonic and ignored by the public," who trusted the BBC over official government programs.[10] Nazi propaganda even failed to generate much liking for the Nazi Party, whose local officials, often incompetent and corrupt, were universally despised.[11] One of the great ironies of these conclusions is that Kershaw reached them partly through an examination of work done by the Sicherheitsdienst (SD), the Nazi intelligence agency. Some of its reports are indeed scathing, such as this note from the SD office in Schweinfurt (a small city of central Germany): "Our propaganda encounters rejection everywhere among the population because it is regarded as wrong and lying."[12]

What about the German armies that fought to the death battles they were certain to lose? Aren't they the ultimate proof of the effectiveness of Nazi propaganda? Studies of German soldiers consistently show that "political values played a very minor part in sustaining their motivation for combat."[13] Instead, as for soldiers everywhere, the main impetus came from the support of the small group they were embedded in, people they had fought alongside for many years, and with whom shared hardships had created special bonds of loyalty.[14] Fear also played a role: fear of being executed if a desertion attempt failed (as thousands of German soldiers were), fear of dying as a POW in the wrong hands (few POWs on the eastern front ever came back, while the relatively lenient treatment of POWs on the western front prompted many desertions).[15]

Kershaw, summing up his findings, notes how "the effectiveness of [Nazi] propaganda ... was heavily dependent on its ability to build on existing consensus, to confirm existing values, to bolster existing prejudices."[16] Whenever propaganda ran against public opinion, it failed abysmally. On the whole, little or no mass persuasion took place. Were the Nazis just particularly bad at propaganda, or do we observe the same pattern in other regimes?

The USSR also made abundant use of propaganda not only during the war but also in the preceding decades, in particular when Stalin was consolidating his power. Early Soviet propaganda efforts failed to resonate with the population. Communist concepts had to be jettisoned in favor of more congenial narratives: patriotism replaced internationalism, the cult of heroes replaced impersonal historical forces.[17] In turn, this strategy backfired badly when many of the heroes were killed during the show trials of the late 1930s. Soviet propaganda never quite recovered. Indeed, even at the apex of Stalinist propaganda, Russian workers and peasants "adopted many tactics of passive resistance," and actively sought out "alternative sources of information."[18] As elsewhere, it is mostly those who benefited from the regime who appeared to accept its values.[19] Even now,

Russian propaganda efforts—for example, in Ukraine—follow a familiar pattern: succeeding modestly when preaching to the choir, backfiring when targeting opponents.[20]

Propaganda by the other great communist power, China, was barely more convincing, even under Mao. Political scientist Shaoguang Wang studied in detail what motivated the various actors of the Cultural Revolution in Wuhan, a large city of central China.[21] Instead of reflecting a "blind faith in Mao," citizens' engagement with the Cultural Revolution was "a product of [their] perception that Mao's initiative would provide solutions to [their] personal problems."[22] Those who stood to benefit from taking up the cause did so; many others resisted.

Work on more recent propaganda attempts by the Chinese government confirms its broad ineffectiveness. A study of Chinese citizens' attitudes toward the government in the mid-1990s revealed that consumption of the state-controlled news media correlated with lack of trust in the government, making it very unlikely that the media successfully instills trust in the leadership.[23] Lack of trust in official media means that Chinese citizens are "always eager to get other information from different channels," as one of them put it.[24] Shortly after Weibo, the Chinese equivalent of Twitter, started up, 70 percent of the Chinese who used social media admitted relying on them as their primary source of information.[25] Mistrust of official media and increased reliance on other sources mean that rumors presenting a negative view of the government are quickly taken up and prove difficult to fight.[26] That Chinese citizens do not passively accept government propaganda is also shown by their many acts of protest. Journalist Evan Osnos, whose work I have drawn on here, reports that, on average, in 2010 there were nearly five hundred "strikes, riots, and other 'mass incidents'" taking place across China *every day*—and that is according to official statistics.[27]

Having had time to learn the limits of propaganda, the Chinese Communist Party has shifted strategies for controlling the public, away from brute persuasion toward what political scientist Margaret Roberts calls "friction and flooding."[28] Friction consists in making sensitive information more difficult to access—blocking keywords, forcing people to use VPNs, or simply not collecting such information in the first place (e.g., information on how well, or poorly, such and such government agency is performing, information that only the state could reliably gather). Flooding consists in distracting people from sensitive issues by bombarding them with official propaganda. The government is suspected of having recruited as many as two million people to spread messages online—known as the 50 Cent Party, after the sum these shills receive per post. Yet, the government seems to have essentially given up on using these legions of propagandists to change people's minds: they "avoid arguing with skeptics . . . and [do] not even discuss controversial issues."[29] Instead, they try to either bolster the views of citizens supportive of the regime in the first place (a significant number, as we will see presently) or talk about other topics, such as celebrity gossip, distracting the attention of citizens who do not care all that much about politics.

In his book on nondemocratic regimes, political scientist Xavier Márquez notes several other failures of propaganda: "Nearly 40 years of Francoist propaganda did not turn Spaniards against democracy . . . constant exposure to the cult of Ceaușescu did not turn most Romanians into his partisans . . . unrelenting propaganda turned many East Germans into habitual cynics who did not believe anything the regime said."[30]

By and large, government propaganda fails to convince the public. It can even backfire, leading to widespread distrust of the regime. At most, propaganda surfs on preexisting opinions and frees people to express what might otherwise be seen as socially objectionable views.[31]

Why, then, do some people in authoritarian regimes behave as if they had been brainwashed, saluting the führer in unison, buying billions of Chairman Mao badges, wailing at Kim Jong-il's funeral? The answer is simple. Every authoritarian regime that relies on propaganda also closely monitors and violently represses signs of dissent. Failure to perform the Nazi salute was perceived as a symbol of "political non-conformism," a potential death sentence.[32] In North Korea, any sign of discontent can send one's entire family to prison camps.[33] Under such threats, we cannot expect people to express their true feelings. Describing his life during the Cultural Revolution, a Chinese doctor remembers how "to survive in China you must reveal nothing to others."[34] Similarly, a North Korean coal miner acknowledged, "I know that our regime is to blame for our situation. My neighbor knows our regime is to blame. But we're not stupid enough to talk about it."[35]

When it comes to genuine support, rather than empty displays, it seems carrots work better than sticks. Chinese citizens might not widely trust the state media, but, on the whole, they respect and support the central government and the Chinese Communist Party—which typically garners more than 70 percent approval, higher than any Western government.[36] It could be propaganda. Or it could be that under the party's direction, China has had high growth rates for decades, lifting eight hundred million people out of poverty.[37]

CAMPAIGNERS

As we have seen repeatedly, forceful attempts at mass persuasion by propagandists in authoritarian regimes fail to sway the population. Rather than the public exercising due vigilance, however, could it be that these failures reflect instead the propagandists' lack of sophistication or skill? Goebbels, for example, didn't seem to have been much of a master influencer: by 1940 already, citizens had lost any interest in official propaganda, on account of its "boring uniformity" (as reported by the Nazis' intelligence service).[38]

The campaign managers, spin doctors, marketers, pollsters, crisis consultants, and other specialists who proliferate in contemporary democracies may be more astute. Authoritarian propaganda relies on monopolistic control of the media: Maybe the lack of competition has blunted the instincts and motivations of the chief propagandists? By contrast, modern political campaigns are fiercely fought, providing plenty of opportunities for professionals to refine their skills and learn how to guide a candidate to victory, as well as for candidates to figure out who can best help them get elected.

I focus here on U.S. politics, for two reasons. First, U.S. politicians vastly outspend other politicians: in 2016, $6.4 billion was spent on political campaigns (a third on the presidential race).[39] Second, that's where the vast majority of studies are conducted.

If the amounts involved are extraordinary, U.S. electoral campaigns—in particular the most high-profile races—are otherwise similar to campaigns elsewhere in being presented by the press as dramatic events, full of plot twists, with candidates going up and down in the polls as a function of devastating ads, moving speeches, and performance in public debates. Indeed, given the means available—the army of volunteers canvasing door to door, the hours of TV ads, the countless robocalls—we would expect commensurately dramatic results.

Yet research on whether political campaigns and the media can win elections, or sway public opinion more generally, has given surprisingly ambiguous results. In the first decades of the twentieth century, a popular model was that of the "hypodermic needle" or the "magic bullet," according to which people would pretty much accept whatever the media were telling them.[40] This model was based on the innovative (but likely ineffective) use of propaganda in World War I, and on a view of the public as reacting reflexively to just about any stimulus they encountered.[41] One thing this model was not based on, though, was data. As opinion polls, tracking of voting behavior, and proper studies of media influence arose, in the 1940s and 1950s, the era of "minimal effects" began.[42] Summing up years of research, Joseph Klapper stated in 1960 that political communication "functions more frequently as an agent of reinforcement than as an agent of change" (a conclusion reminiscent of what the research on propaganda shows).[43]

The 1970s and 1980s saw the rise of experimental studies in political science. Instead of measuring people's opinions in the field, researchers would bring participants into the lab, expose them to various stimuli—campaign materials, TV news, and so forth—and measure the influence of these stimuli on the participants' opinions. These techniques revealed that the media had the potential to influence public opinion: not by telling people what to think but by telling people what to think about (agenda setting), how to best understand issues (framing), and what criteria to use when evaluating politicians (priming).[44] Although these effects are less direct than those suggested by the hypodermic-needle model, they could still be powerful: people who evaluate politicians according to their economic policies rather than their views on abortion (say) are likely to vote differently.

The advantage of these lab-based techniques is the rigor of their methods, as they allow researchers to perform well-controlled experiments, with participants randomly exposed to different stimuli, their reactions carefully monitored. The drawback of these methods is their lack of so-called ecological validity: it is hard to tell whether the phenomena observed in the lab happen in the uncontrolled environment of real life. For instance, some studies showed that exposure to different pieces of news on TV could lead to changes in political opinions. In real life, however, people aren't passively exposed to TV news: they choose which news to watch, or even whether to watch the news at all. Political scientists Kevin Arceneaux and Martin Johnson conducted a series of studies in which participants had greater leeway in which channel to watch. They observed not only that many people simply tuned out but that those people who chose to watch the news were those with the most political knowledge, people who were also less likely to change their minds in reaction to what they saw on the news.[45] Still, even looking at more ecologically valid studies, it is clear that political campaigns and the media can shape public opinion on some issues. But the way they do so reveals that people do not unquestioningly accept whatever message political campaigns put forward.

By far the most important moderator of whether campaigns or the media influence public opinion is the strength of people's prior opinions. On the vast majority of political issues, people have no strong opinion, or even no opinion whatsoever—which makes sense, given the time and effort required to garner information on any topic. For example, in the run-up to the 2000 U.S. presidential election, few voters were aware of what position George W. Bush and Al Gore (the two main candidates) held on Social Security.[46] As a result, when people were told that the candidate from the party they favored had such and such opinion, they tended to adopt this opinion, following "party cues."[47] Following party cues reflects the (largely) sound working of trust mechanisms: if you have come to trust a party over many years, it makes sense to follow its lead on issues about which you have little knowledge. Citizens are also quite skilled at recognizing who among them is most knowledgeable on political issues, and at taking their opinion into account.[48] On the whole, people are more influenced by reliable signals. For example, a newspaper sends a more reliable signal when it endorses a surprising candidate—one who doesn't belong to the party usually supported by the newspaper—and people are only influenced (if at all) by these surprising endorsements.[49]

In the first decade of the twenty-first century, political scientists began conducting large-scale experiments on the effectiveness of political campaigns, sending flyers to a random subset of counties, canvassing a random subset of houses, calling a random subset of potential voters, and so forth. Opinion surveys or voting outcomes were then recorded, allowing the researchers to precisely estimate the effects of their intervention—the letter, the face-to-face discussion, the call—on participants who had been exposed to it, compared with otherwise similar participants who hadn't. This methodology offered the best of both worlds: it was rigorous yet ecologically valid.

In 2018, political scientists Joshua Kalla and David Broockman published a meta-analysis of all the studies that respected these rigorous methods, to which they added some new data of their own.[50] Some of the campaign efforts carried out a long time before the election had a small but significant effect on voting intentions. Early on in the election cycle, people have had less time to develop fixed ideas about who they are going to vote for, making their opinions slightly more labile. However, these effects were never long-lasting and had all but disappeared by Election Day, so that the campaign efforts had no net effect on voting behavior.[51] Other studies have shown some effects of campaign efforts on elections for which voters have few preconceived ideas, as they can't rely on candidates' affiliations, such as primaries or ballot measures.[52]

When it comes to the big prize—voting on congressional or presidential elections—the overall effect of the campaign efforts studied was nil.[53] This is quite a remarkable result. In spite of the huge sums sunk into mailing, canvassing, calling, and advertising, campaign interventions in the most salient elections (in the United States at least) seem, as a rule, to have no effect.

The most recent, sophisticated-looking techniques have done nothing to challenge this conclusion. Many of you will have heard of Cambridge Analytica, the infamous firm that harvested data from Facebook users (often without their consent), created psychological profiles of these users, and offered political campaigns ads targeted specifically to these psychological profiles. According to the *Guardian*, Cambridge Analytica allowed "democracy [to be] highjacked."[54]

In fact, it was a scam.

Targeted advertising can, it seems, have some limited effects, but these have only been proven on product purchases, with relevant data on the users' profiles, and the effects were tiny, adding a few dozen purchases after millions of people had seen the ads.[55] Cambridge Analytica was attempting to influence presidential elections (something no ad has been shown to do) with dubious data on users. Even if the influence of Cambridge Analytica's campaign had been as large as that recorded in experiments on beauty products, it would only have swayed a few thousand voters. In reality, its influence was likely nil. Republican political analysts remember the Cambridge Analytica employees "throwing jargon around," but they never saw "any evidence it worked"—unsurprisingly as it was based, still in their words, "on this pop psychology B.S."[56]

The main explanation that had been previously offered for the inefficiency of political campaigns is that each side invests in reaction to the other's investment, so that their effects cancel each other out. But this cannot be what happens in the studies reviewed by Kalla and Broockman. Who received the treatment—the flyers, the calls, and so forth—was random, making it impossible for the other side to target these people in particular. Political campaigns just don't seem to persuade a significant number of voters, at least in important elections. What about the wide swings in the polls observed throughout the campaigns then? A recent analysis suggests that they are largely artifactual: when a candidate is perceived to be doing well, people inclined to vote for them are more likely to answer the polls, creating the illusion of swings, when in fact few people are changing their minds.[57]

By contrast with most political campaigns, the news media "have an important effect on the outcome of presidential elections," as statistician Andrew Gelman and political scientist Gary King put it more than twenty-five years ago. However, Gelman and King specify that this effect is achieved "not through misleading advertisements, sound bites, or spin doctors, but rather by conveying candidates' positions on important issues."[58] As a rule, the main role played by the media is to provide the information without which citizens couldn't make even minimally informed political decisions, information such as what party each candidate belongs to, or what the candidates' platforms are. Recent research vindicates Gelman and King's pronouncement. More news media coverage makes for a more informed electorate;[59] citizens who trust the media more are also the best informed;[60] more informed electorates are less susceptible to persuasion, not more. As a result, the more news sources are available to the public, the more the public is aware of what politicians are doing, and the more efforts politicians make to fulfill their constituents' wishes.[61] At least in well-publicized elections, the media and political campaigns play a largely positive role of informing citizens—even if, when it comes to American political campaigns, it is easy enough to imagine that the same result could be achieved at a fraction of the cost.

ADVERTISERS

The sums spent on political campaigns pale by comparison with the amounts lavished on advertising. In 2018, more than half a trillion dollars was spent on advertising worldwide.[62] This money can (in theory) heavily influence customers' preferences, making them choose more expensive products, or even favor inferior alternatives—for example, buying Coca-Cola when they are supposed to prefer Pepsi in blind tests.

As for political campaigns, measuring the effects of advertising is difficult. Researchers at Google and Microsoft have argued that to know whether an online ad produces positive returns at all, it has to be tested on more than ten million people—and that is in the ideal scenario of a perfectly controlled experiment.[63] If advertising effectiveness is so difficult to measure, it is not for technical reasons but because ads have small effects at best, making it difficult to tell whether they have any effect at all.

Early work on advertising efficiency suggested that most ads had no discernible effects whatsoever. An early article from 1982 was already asking, "Are you overadvertising?," the answer being a clear yes.[64] A review of studies conducted in the ten years after 1995 claims to have observed some small but significant effects for TV ads.[65] As is the case for political ads, the main variable moderating the effectiveness of consumer advertising is whether or not the audience has preconceived opinions. Ad campaigns have no effect on consumers who have already experienced a product.[66] This result is important, as it means that advertising doesn't function by making some products gain a better image or appear more prestigious—if this were true, people who know the product already should be just as likely to be influenced by the ads. Instead, advertising functions mostly by "giving [customers] information on inherent product characteristics," information that is superseded by personal experience when it is available.[67]

A sad example of advertising working is that of TV ads for cigarettes. Cigarettes aren't exactly a hard sell: long before advertising existed, people had been smoking everywhere tobacco was available. Merely pointing out the existence of cigarettes should be sufficient; the effects of nicotine on the brain, targeting reward centers and soon making itself indispensable, would then do the bulk of the work. As expected, cigarette ads were most efficient when they could tap into a market of people who had not been aware that smoking was an option, such as young Americans in the 1950s.[68]

When advertising affects consumers, it doesn't affect them in a way that reflects sheer credulity. For example, the effectiveness of ads relying on celebrities depends on whether the celebrity is perceived as a trustworthy expert in the relevant domain.[69] By contrast with relevant expertise, gratuitous sex and violence in ads are more likely to decrease their impact.[70]

These results might be difficult to believe: we can all think of celebrities associated with products they have no known expertise in. Boyd and Richerson, defending a bias to do whatever prestigious people do, mentioned Michael Jordan advertising underwear, but a better-known example might be that of George Clooney and Nespresso. As far as I can tell, Clooney has no recognized expertise in coffee, yet he has come to be associated with the brand. However, the direct effects of his endorsement are not clear. Nespresso was already growing more than 30 percent a year before Clooney became its ambassador in Europe, in 2006. In the following years, the brand kept growing at similar rates.[71] Nespresso had also achieved formidable growth in the United States long before Clooney started advertising for the brand there, in 2015.[72] Ironically, if it is unclear how many customers Clooney brought in, we know Clooney was brought in by customers: to reward early buyers for the brand's success, Nespresso asked them to select an ambassador. They picked George Clooney.[73]

Advertising cannot even be blamed for making people choose sodas (supposedly, Coca-Cola) they wouldn't favor in a blind test. Most people are simply incapable of distinguishing between Coke and Pepsi.[74] Even if advertising in this domain had huge effects (which it doesn't), not much persuasion would be involved in making people choose Coke over Pepsi, products essentially indistinguishable by taste or price.

Marketing researcher Gerard Tellis drew from his review of advertising effectiveness these words of caution: "The truth, as many advertisers will quickly admit, is that persuasion is very tough. It is even more difficult to persuade consumers to adopt a new opinion, attitude, or behavior."[75]

PATTERNS OF MASS PERSUASION

As attempts at mass persuasion pile up, from demagogues haranguing crowds on the Agora to advertisers vying for our attention on smartphones, a clear pattern emerges. Mass persuasion is tremendously difficult to achieve. Even the most dreadful propaganda attempts, from Nazi Germany to Stalinist USSR, have been surprisingly ineffective at changing people's minds.

Any message that clashes with our prior beliefs, any injunction to do something we aren't happy to do anyway, is overwhelmingly likely to fall on deaf ears. The Catholic Church at the height of its power could not get peasants to fast, confess, make penance, willingly pay the tithe, or abandon their pagan practices. Nazi propaganda failed to make the Germans abhor the handicapped or like the Nazis. Once voters have their minds set on a candidate, all the campaign money in the world isn't going to sway them. Ads are wasted on consumers who have firsthand experience of the product advertised.

Mass persuasion fails when it encounters resistance. An audience needs to have positive reasons to believe a message if the message is to have any effect. The most effective messages echo the prejudices or serve the goals of their audiences—anti-Semites defending their hatred with Nazi propaganda, revolting crowds recycling millenarian themes—but then can we really talk about persuasion? At best, mass persuasion changes people's minds on issues of little import, as when voters select a party whose platform fits with their opinions on significant issues and then follow the party's lead on less (personally) significant topics.

Clearly, the patterns of mass persuasion aren't compatible with widespread credulity. Instead, they reflect a cautious evaluation of the information communicated, as people decide whether messages fit with their prior opinions, and whether they come from reliable sources.

IF WE ARE SO VIGILANT, WHY ARE SOME MISCONCEPTIONS SO POPULAR?

In these past chapters, I have detailed the functioning of our open vigilance mechanisms: how we decide what is plausible and well argued, where expertise lies, who is trustworthy, and how to react to emotional signals. A wealth of psychological experiments shows these mechanisms function broadly rationally, allowing us to avoid harmful messages and change our minds when confronted with good enough evidence. Open vigilance mechanisms are certainly efficient enough to stop nearly all mass persuasion attempts from changing our minds.

This optimistic conclusion might seem out of touch with the list of patently wrong beliefs—powerful witches are afoot, Barack Obama is a Muslim, vaccines aren't safe—enjoying widespread cultural success. Yet professing a mistaken belief doesn't necessarily make one gullible. In the next six chapters, I explore a laundry list of misconceptions, from rumors to fake news, showing that the ways in which these misconceptions spread, and the effects they have on our thoughts and actions, are best explained by postulating efficient open vigilance mechanisms rather than outright credulity.

10

TITILLATING RUMORS

IN 2015, 20 PERCENT OF AMERICANS believed that Barack Obama, then the sitting U.S. president, had been born abroad. Forty-three percent of Republicans—the opposition party—also thought he was Muslim (Obama was born in Hawaii, a U.S. state, and is a Christian).[1]

In April 2017, David Dao was forcibly removed from an overbooked United Airlines plane. The situation was handled so poorly that he lost a tooth, broke his nose, and got a concussion (according to his lawyer). After Dao and the airline company settled, rumors were flying around the popular Chinese social media platform Weibo that the settlement had reached the sum of $140 million.[2] Although the real amount was never divulged, the likelihood is that it was a hundred times lower.[3]

In early 1969, there appeared in the French provincial town of Orléans a rumor that young women were being abducted from the changing rooms of Jewish retailers, to be sent abroad as prostitutes.[4] In spite of official rebuttals from the police, politicians, and other authority figures, the rumor grew for several months before slowly dying over the summer.

Besides these examples of wildly inaccurate rumors, the low level of accuracy of some rumors is borne out by more systematic studies.

In June 1950, the Indian town of Darjeeling was hit by devastating landslides. Psychologist Durganand Sinha studied the rumors that proliferated in the aftermath—rumors about what had caused the landslides, the number of casualties, the amount of rainfall, and so forth.[5] The rumors were uniformly false, wild exaggerations and dramatizations of the actual events. The same outcome was observed after an earthquake hit the Indian state of Bihar in 1934.[6]

The University of Michigan saw a large strike in 1975. Recognizing that "rumors tend to proliferate in times of crisis," two psychologists, Sandord Weinberg and Ritch Eich, attempted to counteract their spread. They set up a rumor crisis center that employees could call to check the veracity of the rumors they heard through the grapevine. Only about 15 percent of the reported rumors were accurate.[7]

RUMORS OF CRISIS

Why are false rumors so common? After psychologists Gordon Allport and Leo Postman published their influential book *The Psychology of Rumor*, shortly after World War II, most theories of rumor diffusion focused on the state of mind of those who believe and spread rumors.[8] As a review put it, "Rumor generation and transmission results from an optimal combination of personal anxiety, general uncertainty, credulity, and outcome-relevant involvement."[9] According to Allport and Postman's theory, changing environments—a black president in the White House, uncertainty about the outcome of a strike—generate anxiety. Anxiety makes people credulous toward information related to the anxiogenic events. Rumors help people make sense of current happenings, reduce uncertainty about the future, and assuage their anxiety. On top of the anxiety-inducing situation, which would make people more gullible, some individuals are supposed to naturally suffer from a lack of "critical sense," making them particularly good transmitters of rumors, however ludicrous these rumors might be.[10]

These explanations may sound plausible, but they do not fit with the theory put forward here. Uncertainty should make us yearn for certainty, anxiety should make us clamor for reassurance—but only if the certainty and the reassurance are real. Being lulled into a false sense of certainty or security might feel nice, but it is a recipe for disaster. Open vigilance mechanisms should reject messages that we don't have good enough grounds to accept, irrespective of how they make us feel.

The main issue with the theory that people credulously look for anxiety-assuaging rumors is that most rumors are more likely to fuel than to extinguish anxiety.[11] Do we feel safer thinking the local shopkeepers are kidnapping young girls? Do exaggerated claims of disaster-related damages assuage our concerns?

But even if the standard theories can't explain the full pattern of rumor transmission, the transmission of so many false rumors might still challenge my argument that people are not gullible, but are good at evaluating communication. Whether false rumors are anxiety reducing or anxiety increasing, many people accept them, often on the basis of flimsy evidence. This seems like a glaring failure for our mechanisms of open vigilance. But to properly gauge this failure, and to better understand its causes, we must take a look at more efficient cases of rumor transmission.

ALL THAT SPREADS IS NOT FALSE

For many years, the *Wall Street Journal* has published a daily column, "Heard on the Street," which records gossip and rumors flying around the world of finance. In an analysis of this column, economists John Pound and Richard Zeckhauser focused on rumors of takeover attempts—when a company makes a bid to acquire another company.[12] They found that nearly half of these rumors were accurate, making them a valuable source of information, which markets appropriately took into account.[13]

Psychologists Nicholas DiFonzo and Prashant Bordia conducted a series of studies of workplace rumors, collecting nearly three hundred from different businesses—rumors about who was promoted, made redundant, leaving the company, and the like.[14] Although accuracy varied from business to business, it was very high: generally above 80 percent and often 100 percent. For example, these researchers noted, "rumors about who would be laid-off at a large company undergoing a radical downsizing were totally accurate 1 week in advance of formal announcements."[15] These results replicate several older studies of the grapevine in work environments, which all had observed rumor accuracies above 80 percent.[16]

One of these studies looked at a particularly interesting environment: the military during World War II.[17] By contrast with the classic study by Allport and Postman, which mostly looked at wartime rumors among U.S. civilians, psychologist Theodore Caplow focused on rumors circulating in the U.S. Army—who was going to be deployed where and when, who would be repatriated, and so forth.[18] These rumors were uncannily correct. According to Caplow: "Every major operation, change of station, and important administrative change was accurately reported by rumor before any official announcement had been made."[19]

Some of the accurate rumors reviewed here might have made people less anxious: soldiers hearing they would soon be repatriated, employees discovering they would get promoted. Undoubtedly, others generated significant stress: hearing one would be sent to the front or made redundant. Whether a rumor increases or reduces anxiety has little to do with its accuracy. What is special, then, about the contexts that consistently generate accurate rumors?

SPONTANEOUS RUMOR TRACKING

At heart, the answer is quite simple: rumors tend to be accurate when their content has significant consequences for the people among whom they circulate.

Like any other cognitive activity, open vigilance is costly, and we only exercise it to the extent that it is deemed worthwhile.[20] This means that in domains that matter to us, we carefully keep track of who said what, and whether what they said turned out to be correct or not. In turn, this motivates speakers to exercise great caution when reporting rumors, so as not to jeopardize their own credibility.[21] When we find out, eventually, whether the rumors were true or not, our ability to track who said what helps us create networks of reliable informants.

This is what enabled the U.S. soldiers studied by Caplow to be so efficient at transmitting accurate, and only accurate, rumors.[22] Given the content of the rumors—such as when and where one would be deployed—it soon became clear whether they had been true or not. Thanks to repeated feedback, the soldiers learned who they could trust for what type of information, and who should be taken out of the information network.

Moreover, for issues that relate to their immediate environment, people are generally able to check the content of rumors, either against their existing knowledge or by gathering new information. This nips false rumors in the bud, irrespective of how anxiogenic the situation might be.

Psychologist James Diggory studied the rumors that surrounded an outbreak of rabies in 1952 in eastern Pennsylvania.[23] People in the most affected counties would have been the most anxious. However, they were also less likely, compared with those in more distant counties, to believe in exaggerated rumors about the threat. The proximity of the threat made them more anxious but also put them in a better position to evaluate the risks accurately.

One of the most poisonous rumors circulating in the United States during World War II accused individuals of Japanese ancestry of treason, in particular of having engaged in acts of sabotage by assisting the attack on Pearl Harbor. While these rumors ran wild in the mainland, in Hawaii, where the suspected individuals lived, they were roundly rejected "for the people could see for themselves and could talk to the various defenders of the islands."[24]

Sometimes, new problems arise that are practically relevant, but about which we don't know much, and reliable networks of informants haven't had time to crystallize. This is likely what happened during the strike at the University of Michigan. In this novel situation, few employees had reliable information about important matters—whether classes would be canceled, whether penalties would be imposed for striking, and so forth. The lack of reliable prior knowledge or established networks created a rich breeding ground for false rumors. However, because the issues were practically important for the employees, they made use of the crisis call center created by the researchers. As a result, "in most cases, false rumors were quelled before they could be widely disseminated."[25]

HOW DO WE BELIEVE IN FALSE RUMORS?

Clearly, our mechanisms of open vigilance can do a very good job when we're assessing the majority of rumors, especially those that affect us most. Why, then, do they seem to fail so abysmally in other cases? I argue that the diffusion of false rumors isn't as much of an indictment of our open vigilance mechanisms as it seems—in fact, quite the contrary.

What is shocking when it comes to false rumors is that people accept them on the basis of such flimsy evidence. But how do people really believe in these rumors? Believing something—a rumor or anything else—is not an all-or-nothing matter. Believing depends on what you do with a given piece of information. A belief can remain essentially inert, insulated from cognitive or behavioral consequences, if we don't work out what inferences or actions follow from it. Dan Sperber has called such beliefs reflective, by contrast with intuitive beliefs, from which we freely draw inferences, and which we spontaneously use in grounding our actions.[26] For example, you intuitively believe there's a book (or other device) in front of you when you're reading these lines. You can grasp the book, you know you can use it to cover your face from the sun, that you can lend it to a friend, and so on. By contrast, take the belief that most stars you can see at night are larger than the sun. You should be genuinely persuaded it is true, and yet there isn't much you can do with it.

For reflective beliefs—beliefs that tend to have fewer personal consequences—we shouldn't expect open vigilance mechanisms to make as much of an effort: Why bother, if the belief doesn't make much of a difference? I argue that most false rumors are held only reflectively, for they would have much more serious consequences if they were held intuitively.

In some cases, it is difficult to imagine what significant behaviors could follow from a rumor. Chinese citizens are hardly going to challenge the way insurance settlements are handled in the United States. A Pakistani shopkeeper might say the Israelis orchestrated 9/11, but what is he going to do about it?

Even when people could do something on the basis of a (false) rumor, they most often don't. American truthers—who believe 9/11 was an inside job—don't act as if they intuitively believed in the conspiracy. As journalist Jonathan Kay noted: "One of the great ironies of the Truth movement is that its activists typically hold their meetings in large, unsecured locations such as college auditoriums—even as they insist that government agents will stop at nothing to protect their conspiracy for world domination from discovery."[27]

Or take the *rumeur d'Orléans*, which accused Jewish shopkeepers of kidnapping young women. Many of the town's inhabitants spread the rumor, although for the vast majority of them, the rumor had little or no behavioral consequences. Some young girls started visiting other retailers, or asked friends to accompany them while shopping in the suspect stores. At the height of the rumor, some people in the busy streets stopped and stared at the shops. Glaring is hardly an appropriate way to react after accusations of submitting young women to a lifetime of sexual exploitation. These behaviors (or lack thereof) show that most of those who spread the rumor didn't intuitively believe in them.

By contrast, the rumors circulating in the wake of Pearl Harbor against Americans of Japanese ancestry seem to have had significant effects, as the U.S. government decided to detain most of these citizens in internment camps. In reality, there were more important drivers behind the internment camps than the nasty rumors about treason. Many of these Japanese Americans had been successful farmers in California, with more productive plots than their white neighbors. Their success led to a "resentment from white West Coast farmers," which "provided part of the impetus for mass incarceration of [Americans of] Japanese descent."[28]

The lack of action following acceptance of the false rumors described here suggests that open vigilance mechanisms barely gave these rumors passing grades. If our open vigilance mechanisms had really deemed the rumors plausible, we should have expected altogether more powerful reactions, the kinds of reactions we witness when people intuitively believe in rumors.

In Pakistan, conspiracy theories about the dreaded ISI—the intelligence service—are very common. Yet Pakistanis don't organize conferences on how evil and powerful the ISI is. Precisely because they intuitively believe the ISI is evil and powerful, they don't say so publicly.

Imagine that a female friend runs out of a shop in tears, crying that she has been the victim of a kidnapping attempt. Will you be content with glaring at the vendor and, later, telling other people to avoid the shop? Aren't you instead going to call the police immediately?

The fact that most people don't take false rumors or conspiracy theories to their logical conclusion is also driven home by the few individuals who do. Edgar Maddison Welch was one of them. He believed the rumors saying that the basement of the Comet Ping Pong restaurant was used by Hillary Clinton cronies to engage in child sex trafficking. Given this belief, coupled with his mistrust of the corrupt police, Welch's storming of the restaurant, guns ablaze, requesting the owners to free the children, kind of made sense. Most people who endorsed the rumor—and, according to some polls, millions did—were happy doing nothing about it or, at worst, sending insulting messages online.[29] One can hardly imagine a child sex trafficker coming to see the error of his ways as a result of reading Nation Pride's commenting on the trafficker's "absolutely disgusting" behavior and giving his restaurant only one star (Google review might want to offer the option of giving no stars for pedophile-friendly pizzerias).[30]

Why did Welch take the pizzagate rumors so seriously? I honestly don't know. What matters for my argument is that of the millions of people who believed the rumor, he was the only one to act as if he did so intuitively.

UNFETTERED CURIOSITY

Even if false rumors do not, as a rule, have any serious behavioral consequences, many people endorse them. Isn't that a failure of open vigilance, even if a more modest one? To understand why this might not be a significant failure, and why people say they believe false rumors, we must start by asking why people are interested in such rumors at all. After all, if they don't do much with the information, why are people so keen on hearing and spreading rumors?

Cognition is costly—a small cost for each bit of information processing, and a substantial cost for growing the brain that enables it all. As a result, our minds are particularly attuned to useful information. We come equipped with a mechanism to recognize human faces, but not human necks.[31] We are naturally attentive to many features of potential romantic partners, but not of programming languages. We are more interested in information about individual humans than individual rocks.

Ideally, we should only pay attention to, process, and store information that is of practical importance, information that allows us to better navigate the world. However, it is impossible to anticipate exactly which piece of information will come in handy—indeed, attempting to make such guesses is also a cognitively costly task. Your friend Aisha bores you with trivial details about her new colleague, Salma. But if you later meet Salma and get a crush on her, this information might come in handy. Processing and memorizing information is costly, but ignoring information can be costlier, so it makes sense to err on the side of caution, especially if the information fits a template of information that is particularly costly to ignore.

Take face recognition. Our ability to recognize faces evolved because it helped us interact with other people. The focus, or proper domain—to use Dan Sperber's terminology—of face-recognition is composed of the faces of actual humans with whom we interact.[32] But a great many objects are picked up by our face recognition mechanism, even though they aren't in this mechanism's proper domain: nonhuman animal faces, a mountain on Mars, electric sockets, and so forth (figure 3; Google "pareidolia" for many more examples).[33] This constitutes the actual domain of the face-recognition mechanism: all the things it can take as input.

Why is the actual domain of our face-recognition mechanism so much broader than its proper domain? Because of a cost asymmetry. If you see a face in an electric socket, your friend might find that funny; if you mistake your friend's face for an electric socket (or anything else, really), she will be significantly less amused.

Mismatches between the proper and the actual domain of cognitive mechanisms create vast domains of relevance: information we find relevant irrespective of whether it has any practical consequence. This is the source of our boundless curiosity.

Like representations of faces, most cultural products are successful because we find them relevant. Celebrity gossip is an example. If information about other people tends to be valuable, information about popular, beautiful, strong, smart, dominant individuals is even more valuable. During our evolution, we would barely have heard of such individuals without actually

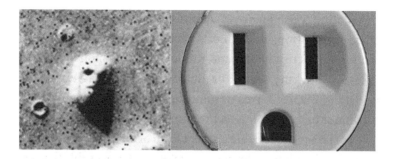

FIGURE 3. Two examples of pareidolia: seeing faces where there are none. *Source:* NASA and grendelkhan.

interacting with them: most of the relevant information was also practically relevant. Nowadays, we may never interact with these salient individuals, yet we still find information about them alluring. If you aren't into the latest gossip about Prince Harry and Meghan Markle, you might be interested in biographies of Lincoln or Einstein, even though you're even less likely to meet them than are people who read *Star* to meet Harry and Meghan. Because of our interest in information about salient individuals, we reward individuals who provide such information, by thinking a friend who knows the latest celebrity stories more entertaining or by buying an author's books.

Many successful false rumors are about threats. It might seem curious that we like thinking about threats, but it makes sense. We may not like threats, but if there are threats, we want to know about them. Even more than faces, information about threats presents a clear cost asymmetry: ignoring information about potential threats can be vastly costlier than paying too much attention to such information. This is true even when the threats are reported in rumors. Nearly a year before Pearl Harbor, the U.S. ambassador to Japan heard that plans for an attack were being hatched, but he dismissed the rumors as unreliable, with devastating consequences.[34] As a result of these costs asymmetries, information about threats is often deemed relevant even if it is not practically relevant. Rumors about the toll of natural disasters, lurking sexual predators, or conspiracies in our midst are bizarre forms of mind candy: guilty pleasures that might not be good for us, yet we can't help but enjoy.[35]

Conspiracy theories are a salient form of threat. Given the importance of coalitions during our evolution, it is plausible that we could have evolved to be particularly attuned to the risk raised by an alliance forming against us.[36] Even if we do not have anything like a dedicated "conspiracy detector," conspiracy theories combine elements that make them relevant: they are about a coalition (jackpot) of powerful people (double jackpot) who represent a significant threat to us (triple jackpot).

A study that looked at more than a hundred thousand rumors on Twitter found that the most successful false rumors (compared with true rumors) were those that elicited disgust and surprise.[37] The logic of cost asymmetry applies to disgust: as a rule, it is better to find too many things disgusting (thereby avoiding potential pathogens) than too few. As for surprise, it is simply a general measure of relevance: everything else being equal, more surprising information is more relevant information. Our five- and seven-year-old sons encapsulated this wisdom while they were discussing which pieces of rubbish found on the beach they liked most: "the things that are disgusting, and that we don't know, that's what's interesting."

By this logic, times of crisis are prone to rumormongering not because crises make people more gullible but because they make people curious about topics they had no interest in previously, from the amount of rainfall to a rabies epidemic on the other side of the state. These new sources of relevance are often not matched by an increase in practical relevance, so that false rumors can spread with relative ease. By contrast, if the rumors have serious practical consequences, people do their best to check them, whether or not there is a crisis situation.

REWARDING RUMOR RELAYS

We tend to reward people who provide us with relevant information: we like them more and think of them as more competent and more helpful.[38] In order to win as many of these rewards as possible, we should be able to ascertain the social relevance of a piece of information—how valuable it is for other people—so that we know what to transmit.

Sometimes this means recognizing that a piece of information would be relevant for a specific individual: if you know a friend of yours is a big Lego fan, it is good to realize information about a Lego exhibit would be relevant for her. To compute this narrow social relevance, we rely on our knowledge of the preferences and beliefs of specific individuals.

In other cases, we're more interested in figuring out whether a piece of information would be relevant for many people. To compute this broad social relevance, we use our own mind as a guide: information that happens to be relevant for us is deemed relevant for others (whether or not it has any practical consequences). This isn't a trivial process. When nonhuman animals encounter information that triggers many inferences—the trail of a prey, signs of a predator—they don't think to themselves, "Damn, that's interesting!" They just perform the required operations and engage in the appropriate behavior. Humans, by contrast, are able to represent the relevance of the stimuli they encounter. When we hear something shocking on the news, we don't simply adjust our beliefs; we also take note of this piece of information's broad relevance, so we can share it later. The same goes for jokes, stories, tips—and rumors. When a rumor taps into the actual domains of many cognitive mechanisms, such as those related to threats or conspiracies, we realize that the rumor is likely to have broad social relevance.

Information that has broad social relevance has a special property: it becomes more valuable because of its own relevance.

As a rule, you score social points when you give people information they find relevant. The friend you told about the Lego exhibit will note that you are thoughtful and well informed, because this information was useful to her. But what if this information were useful not just to her but to other people she knows? Then she would, in turn, be able to score social points by spreading this information. When we give people information that has broad social relevance, we score double points: points because they find the information relevant, and more points because they can use the information to score points in turn, and they are thankful for it.

News that we get through mass media often has broad social relevance (that's why the media are talking about it), but because the media have a wide reach, it can be difficult to make much social use of them. By contrast, rumors are the perfect material to appear interesting: because they typically spread to only one or a handful of individuals at each step, they provide us with plenty of opportunities to score social points—not only because our interlocutors find the rumor relevant but also because it will help them score points in turn.

MINIMAL PLAUSIBILITY

Rumors tap into the actual domain of many cognitive mechanisms, giving them the potential to be highly relevant. But to be genuinely relevant, they also have to be plausible. Some communicated content, from jokes to fairy tales, can spread successfully without being considered true, or even plausible. This is not the case for rumors. "Children are being sexually abused in a restaurant!" is neither funny nor entertaining. If it is not at least somewhat plausible, it is completely uninteresting.

If we like rumors in part because they allow us to score social points by spreading them, we should be careful that we don't spread implausible rumors. Or, more precisely, we should be careful that we don't spread rumors that others find implausible. If we do, we not only fail to gain any benefit but also might suffer some costs as a result: as a rule, people will not trust us as much if we provide them with information deemed implausible.

As I have argued in chapters 3 to 7, when it comes to evaluating what others tell us, open vigilance mechanisms are mostly on the lookout for cues that the message should be accepted. Absent such cues, the default is rejection. Given the risks of communication—accepting the wrong piece of information could be life-threatening—this is a safe and sensible way to operate.

By contrast, when guessing whether someone else is going to accept or reject a piece of information, the potential costs are lower. In the vast majority of situations, the risk of saying something wrong is being thought a little less smart, a little less diligent. Only when people suspect duplicity, and when the stakes are high, could the costs be prohibitive—and this is not the case with most false rumors. These social costs should not be neglected, but they are much lower than the potential costs of being misled by others. As a result, when attempting to estimate how others evaluate our messages, we can use the opposite strategy to that used when we evaluate theirs: look for cues the message might be rejected, rather than cues it will be accepted. In the absence of any indication that our interlocutors are predisposed to reject the message, our default is to think they will accept it.

The way people spread the *rumeur d'Orléans* fits well with this indirect use of open vigilance mechanisms. People seemed to have no qualms about spreading the rumor far and wide. Only when the citizens of Orléans thought someone would be in a good position to reject the rumor did they choose to not share it with them. No one directly shared the rumor with the town's older, better integrated Jewish population, who might have personally known the suspected shopkeepers, and thus be in a position to refute the rumor. Likewise, the only people to call the police were those who genuinely wondered whether the rumor was true

Counterintuitively, open vigilance mechanisms are actually doing their job when it comes to false rumors. First, they evaluate the rumor to decide whether it should be accepted, and their verdict is: not really. This leads to a merely reflective acceptance, precluding costly behavioral consequences. Second, the same mechanisms are used in a roundabout manner, to gauge the odds that others would reject the rumor, helping us avoid the social costs of sharing the rumor with those who might deem it ridiculous, and thus judge us negatively.

ESCAPE FROM REALITY

Rumors take different forms and circulate in different ecosystems. At one extreme we find rumors that are practically relevant for at least some of those involved—who will get fired or promoted, sent to the front or repatriated. This practical relevance motivates people to circulate the information, as they gain social points for doing so. But it also makes them conscious that the information they provide will be checked by knowledgeable individuals and, if found wanting, damage the reputation of those who shared it. Rumors that circulate in this way are overwhelmingly likely to be accurate.

At the other extreme we find rumors that are of little practical consequence, but that people think have broad social relevance. These mind-candy rumors titillate our interest in information about famous people, threats, conspiracies, and so forth. The fact that they are mind candy has several consequences. It means not only that we are interested in them but also that we expect others to be interested in them. We even expect our interlocutors to be interested in them because their interlocutors will be interested in them. By spreading a juicy rumor, we give people an opportunity to spread it further, for which we get some kudos.

Several factors make it difficult for these rumors to be corrected. The lack of practical consequences means that the beliefs don't properly interact with the actual world, traveling in a world of expectations about what other people find interesting that remains one step removed from reality.[39] As rumors and reality come into contact, those who hold false rumors are promptly disabused. When Edgar Maddison Welch acted on the basis of the rumors about Comet Ping Pong, he quickly suffered the consequences. Had more people done so, the rumor would never have had a chance to spread.

Moreover, not only the benefits but also the costs of sharing these rumors are social. To avoid these costs, we withhold the rumors from those most likely to prove skeptical, further lowering the chances of getting negative feedback. One of the reasons the *rumeur d'Orléans* proved so successful is that the inhabitants studiously avoided disclosing the rumor to policemen or to anyone who might know better, further reducing the chances that the rumor interacts with reality.

Conspiracy theories go even further in hampering good feedback, as the people in the best position to know whether there is a conspiracy are those accused of engaging in the conspiracy. But, obviously, if the conspiracy were true, the conspirators wouldn't admit to it, making any denial suspicious. Again, the *rumeur d'Orléans* provides an example of how easy it is for rumors to turn into accusations of conspiracy, and even for the fight against rumors to take a conspiracist turn. In its initial form, the rumor wasn't a standard conspiracy theory, because the presumed evildoers—the shopkeepers accused of kidnapping girls—weren't in a position of power. But as the rumor grew, it started facing contradictions: If these crimes were widely known, why weren't the police doing anything? They must have been paid off. The politicians who attempted to debunk the rumor must also be in on it. It is only then that the rumor was ready to collapse under its own weight: too many people knew local policemen or politicians for the rumor to be credible any longer, at least for the majority of the population.

WHAT TO DO?

If false rumors spread so well, it isn't because people take them too seriously but because they don't take them seriously enough.

To increase the ratio of true to false rumors, then, we should try to close the gap between social and practical relevance. When we find a rumor appealing, we should pause before spreading it further—by gossiping to a friend or hitting retweet. What would we do if we had to make a practical decision based on this rumor? Would we engage in vigilante justice to stop the sexual abuse of children? Thinking of a rumor in practical terms should motivate us to check it further.

Some information, however, can hardly be made to have practical consequences. For instance, the exaggerated stories that appear in the wake of a natural disaster are hardly actionable. In these cases, we can at least imagine sharing the information with someone who would be in a good position to know, or who would be affected by the information (which is often the same thing). Many would-be truthers would balk at the idea of sharing their suspicions with 9/11 first responders. Someone who happily ventures the figure of $140 million as a settlement for David Dao's misfortune to shock his friends might fear looking stupid in front of an experienced personal injury lawyer.

Imagining the personal costs of acting on the basis of a rumor is a good first step in the fight against false rumors.[40] At least, we're less likely to be part of the problem. To be part of the solution, we should try to inflict some costs on those who spread false rumors—at the very least deny them any benefits. We shouldn't hesitate to raise doubts, to question the plausibility of their stories or the reliability of their sources (the importance of which I will explore in the next chapter). This typically entails a social cost. After all, people like a juicy rumor, which they can use to score social points. We don't thank the skeptic who spoils it for everyone. To reduce this cost, and to increase the effectiveness of our questioning, we should be as polite as possible, restrain from imputing nefarious intentions to those who spread questionable rumors, and be careful not to overstate our own claims. Are we really sure the rumor is false?

We should think of these personal costs as a contribution to the public good. We would all be a little bit better off if fewer false rumors circulated. However, to reach such an equilibrium, many people will have to deprive themselves of potential benefits, refusing to spread juicy rumors and incurring the small cost of being thought a skeptical killjoy.

11

FROM CIRCULAR REPORTING TO SUPERNATURAL BELIEFS

ONE OF THE MAJOR FACTORS that distinguishes accurate and inaccurate rumors, and that I haven't touched on yet, is the quality of sourcing. By sourcing I mean providing our interlocutors with a description of how we got a piece of information.

In environments conducive to the spread of accurate rumors, people say things like "Bill Smith from HQ said that John was going to be repatriated."[1] Here, the speaker identifies the relevant source of information, allowing his interlocutor to gauge its accuracy more easily. Proper credit (if the information turns out to be true) or blame (if it doesn't) can be given not only to the speaker but also to Bill Smith. This motivates members of the network to be more careful when starting rumors, as false rumors jeopardize their reputation not only with those with whom they share the rumor but also with every individual to whom the rumor then spreads.

By contrast, inaccurate rumors are accompanied by vague sourcing ("people say that . . .") or, worse, by inaccurate sourcing that increases the credibility of the rumor. The *rumeur d'Orléans* was made more plausible by its sources: "A friend's father is a cop, and he's investigating a kidnapping case . . ." or "My cousin's wife is a nurse, and she treated the victim of an attempted kidnapping . . ."[2] The obvious problem with these sources is that they are simply false: no such cop or nurse exists. A less obvious problem is that the sources remain the same throughout the chain of rumor transmission.

In theory, the information about the (imaginary) credible source should have become increasingly diluted as the rumor was passed along, going from, say, "a friend's father" to "a friend's friend's father," "a friend's friend's friend's father," and so forth. But this is not what the researchers observed. The length of the chain was never acknowledged. Instead, most of the people reported the rumor as being validated by "a friend's father" (or the equivalent cousin's wife, etc.). As sociologist Edgard Morin, who led the team studying the *rumeur d'Orléans*, put it: "Each new transmitter [of the rumor] suppresses the new link, and rebuilds a chain with only two or three links."[3]

Sourcing can greatly help or hinder the work of open vigilance mechanisms. Why does it work so well in some cases, and so poorly in others? To better understand, we must start by appreciating the omnipresence of sources.

OMNIPRESENT SOURCES

Paying close attention to sources might seem to be the remit of professionals. Since Thucydides and his *History of the Peloponnesian War*, historians have reflected on which sources their work should be based on, distinguishing primary from secondary sources, debating the reliability and independence of their sources—engaging in historiography. More recently, journalists have also learned to practice source criticism, not relying on a single source, finding independent means of evaluating their sources' credibility, double-checking everything. Clearly, in some domains, people must take great care to find, track, evaluate, and cross-check their sources. Absent this learned, reflective practice, the information provided by academics or journalists cannot be relied upon.

Yet sourcing isn't restricted to professionals. We all do it, all the time, but usually in an intuitive, rather than reflective, manner. For instance, imagine asking your friend Aluna about a movie you consider watching. She tells you one of the following:

(1) It's good.
(2) I saw it last week, it's good.
(3) I heard it's good.
(4) Osogo told me it's good.
(5) The *Chicago Sun-Times* says it's good.

Even though the opinion ("it's good") remains the same, you weigh it differently as a function of how it is presented. The least convincing would likely be (3), because it provides little information with which to evaluate the opinion. How you weigh the others would depend on your judgment on the tastes of Aluna, Osogo, and the *Chicago Sun-Times* movie critic. Even when no source is explicitly provided, as in (1), you would likely be able to draw some inferences: if Aluna utters (1), she is more likely to have seen the movie than to base her opinion only on the trailer, or on a movie review. The provision of information about sources gives more fodder to our mechanisms of open vigilance.

Specifying the source of our statements is so important that many languages make it grammatically mandatory. In English, you must indicate the tense of the verb to make a grammatically correct sentence. In Wanka Quechua, a language spoken in the south of Peru, you must specify how you acquired a given piece of information:

(1) Chay-chruu-**mi** achka wamla-pis walashr-pis: "Many girls and boys were swimming" (**I saw them**)

(2) Daañu pawa-shra-si ka-ya-n-**chr**-ari: "It (the field) might be completely destroyed" (**I infer**)
(3) Ancha-p-**shi** wa'a-chi-nki wamla-a-ta: "You make my daughter cry too much" (**they tell me**)[4]

The bits in bold—the evidentials—are used to tell whether the speaker owes their belief to direct perception, inference, or hearsay. Wanka Quechua is far from being unique: at least a quarter of the world's languages possess some kind of evidentials.[5] Some languages have relatively simple systems with only two evidentials, such as Cherokee, which distinguishes between firsthand and non-firsthand information.[6] Other languages, like Kaluli, spoken in Papua New Guinea, have complex systems, with a great variety of evidentials to choose from.

Whether they are conveyed through evidentials, with explicit mentions ("Peter told me"), or left implicit ("This movie is good" suggests direct experience), sources are omnipresent in language. Why?

For open vigilance mechanisms, an obvious role of source information is to make a statement more convincing, for example, by specifying that it stems from direct perception or from a reliable individual. But why would sources make a statement more convincing? After all, if you don't trust the speaker enough to accept their statement ("Paula is pregnant"), it's not immediately obvious why you should trust them when they give you source information ("I've seen Paula"). Indeed, when the speaker is thought to be dishonest, source information doesn't help. If your poker partner tells you, "I'm looking at my hand, I have a royal flush, you should fold," they won't be any more convincing than if they told you, "You should fold."

Fortunately, in most interactions we don't suspect our interlocutors of such dishonesty. But that doesn't mean we trust them entirely; far from it. In many cases we don't think them competent or diligent enough to warrant changing our minds only on the basis of their opinion. It is in these situations that providing sources makes statements more convincing. For example, I trust my wife a lot. I trust her with our kids. I would trust her with my life. But if, while we're shopping, I think there are eggs in the fridge and she says there aren't, I don't believe her. It's obviously not that I suspect her of lying but that I don't have reasons to believe she's in a better position than I am to know what the egg situation is. If she tells me, "I checked the fridge before leaving; there were no eggs left," I'm convinced that we have to buy more eggs.

By default, statements are attributed to the speaker's ability to draw inferences, which becomes the main locus for estimations of competence: we believe people more if we think them better at drawing inferences (in the relevant domain). By providing sources we outsource (pardon the pun) competence to other cognitive mechanisms—chiefly, perception—or to other people. These other sources become the locus of the estimation of competence and can be used to convince our interlocutors when they believe that our senses, or some third parties, are more reliable than our inferential abilities.

Yet people do not always provide information about the sources of their beliefs in order to persuade their interlocutors. Sometimes, source information has the opposite effect. If Bill says, "Someone told me Paula is pregnant," you might be less inclined to believe him than if he had simply said, "Paula is pregnant." Why would Bill give you reasons not to believe him?

We use what we know about our interlocutors—how competent and diligent they are—to evaluate their messages, but we also use what we know about messages to evaluate our interlocutors. If we know for sure, or later find out, that Paula isn't pregnant, our opinion of Bill as a competent and diligent source of information decreases. But it decreases less if he hedged his statement ("Someone told me that Paula is pregnant") than if he took full responsibility ("Paula is pregnant").[7]

Conversely, Bill might want to get more credit than he deserves for an idea by obscuring its actual source.[8] If he tells you, "I think Paula is pregnant," and you knew that Paula wasn't pregnant three months ago, you might attribute to him the ability to recognize early pregnancy based on subtle cues, and maybe some more general social skills. But if he tells you, "Paula told me that she's pregnant," he won't get much credit: he just had to listen to what Paula was saying.[9]

TWO DEGREES OF SEPARATION

Providing information about sources serves two broad functions: to convince our interlocutors, and to engage in reputation management (i.e., to take more credit than warranted or, on the contrary, to limit our exposure to reputational fallouts). The interplay of these goals helps explain inaccurate sourcing and its effects.

Take the sources that often accompanied the *rumeur d'Orléans* ("A friend's father is a cop . . . ," etc.). Why did speakers provide such sources? And why did their interlocutors accept them?

For those who provide them, these sources play a dual role: to increase credibility, and to limit exposure if their interlocutor questions the validity of the rumor. However, it seems as if this second goal contradicts the thesis I have defended in the last chapter, namely, that people spread rumors mostly so that they can score social points: How could they both not be exposed if things go wrong (the rumor is rejected), and get credit if things go well (the rumor is accepted)? Those who spread rumors can achieve this apparently impossible feat because the value of such wild rumors isn't so much in the practical implications of their content as in being something people want to hear about—a type of mind candy. So someone who transmits the rumor can manage both to keep their distance from the content (they don't say they have witnessed anything themselves) and to get credit for giving their interlocutors a gift they can use to score social points in turn.

From the speakers' point of view, an external, somewhat credible source hits a sweet spot as a way of increasing plausibility, decreasing exposure, and improving the overall credit they can obtain from spreading the rumor.

We see the same pattern in the spread of different types of false information. In the 1980s a fear formed in the United States that snuff movies (movies depicting actual murders, torture, or rape) were being regularly shot and widely distributed. Nearly all the people who spread these accusations remained at least one step removed from any actual witnesses: they had never seen a snuff movie themselves, but they knew someone who had.[10]

Similarly, advocates of conspiracy theories rarely rely on firsthand knowledge. Few people claim to have witnessed the shooting by Stanley Kubrick of the fake moon landing, or to know the guy who actually killed JFK. Not even David Icke says he has seen with his own eyes the human-reptilian beings that control the earth. Instead, he claims to have "started coming across people who told [him] they had seen people change into a non-human form."[11]

HIDDEN DEPENDENCIES

Before launching the second Iraq War in 2003, the administration of President George W. Bush engaged in a vast program of justification. One of its key arguments, used by Bush as well as by high-ranking officials such as Condoleezza Rice and Colin Powell, was that the Iraqis had attempted to buy "significant quantities of uranium from Africa."[12] Several intelligence agencies, across at least two countries—the United States and the United Kingdom—said they had documents in their possession proving Saddam Hussein's attempt to buy from Niger hundreds of tons of uranium oxide, a material that can be processed and used to build nuclear weapons. This convergence of reliable sources—the Central Intelligence Agency (CIA), the Defense Intelligence Agency (DIA), the UK intelligence services—allowed these accusations to play a central role in the justification for the war.

In fact, the evidence all rested on documents peddled by a former Italian spy to several intelligence agencies. The concordance between the agencies' assessment was only as good as this set of documents. And the documents were straight-up forgeries. Not only had Hussein not attempted to buy uranium from anybody, but he had given up on his nuclear weapons program more than ten years earlier, in 1991.

These forged documents wreaked such havoc in part because of the White House's eagerness to justify the war, but also because the intelligence agencies failed to disclose their sources. The British agencies in particular played an important role, as they were seen as providing more independent evidence than the various American agencies. But they never revealed to their U.S. colleagues the basis for their accusations, depriving them of the ability to work out that all the evidence came from the same set of documents. As a U.S. intelligence official quoted by the *Los Angeles Times* put it: "This became a classic case of circular reporting. It seemed like we were hearing it from lots of places. People didn't realize it was the same bad information coming in different doors."[13]

Let's forget for a minute the documents were forgeries, as this is not the aspect of the story I'm interested in here. If the documents were real, for each individual agency, disclosing the source would make its case more convincing. However, given that all the agencies relied on the same source, their case was more convincing when the sources weren't disclosed and their opinions were thought to have been formed independently of each other. As explained in chapter 5, a convergence of opinions is a reliable indicator of the opinions' validity only to the extent that the opinions have been formed independently of each other. If they all rely on the same source, they are only as strong as the one source.[14] In this case, the combined agencies' case would have actually been less convincing had they disclosed their sources—even though doing so would have made each individual case seem more convincing.

When the agencies failed to reveal their sources, there was a hidden dependency between their opinions. Such hidden dependencies are a particularly tricky problem for our mechanisms of open vigilance. For each informant—here, an intelligence agency, but the same applies to any other case—their statements are made less convincing by the absence of a source. As a result, our mechanisms of open vigilance have no reason to be on the alert: they are on the lookout for attempts to change our mind, not attempts *not* to change it. When someone fails to mention a source that would make their statement more convincing, we're not particularly vigilant. If many people do the same thing, we might end up accepting all of their statements, without realizing they all stem from the same source, ending up more convinced than we should be. Not identifying hidden dependencies is one of the rare failures of open vigilance mechanisms that lead to the acceptance, rather than the rejection, of too many messages.

Other, more mundane situations give rise to hidden dependencies, situations in which many speakers appear to have come to an agreement independently of each other, when in fact their opinions largely stem from the same sources.

WHY DO BELIEVERS SAY THEY BELIEVE?

We live in wonderful times: the 14th Dalai Lama is on Twitter. His message is one of peace and toleration, including toward other religions: "Because of the great differences in our ways of thinking, it is inevitable that we have different religions and faiths."[15] This is an appealing view: we each find the religion that best suits our needs and frame of mind. However inspirational this statement might be, it is quite obviously false. For the vast majority of people today, and nearly all of our ancestors, the religion they adopt is not determined by their own way of thinking but by where they happen to be born. People born in isolated Amazonian tribes rarely develop a spontaneous belief in transubstantiation. People born in rural Pennsylvania don't tend to grow up believing in reincarnation.

Beliefs that are specific to different religions—such as transubstantiation—are pretty clearly socially transmitted. Indeed, for most of history, just about everyone adopted a version of religion that was quite similar to that of their elders. (Note that here I'm using a very broad definition of religious beliefs, that encompasses beliefs in the supernatural, in creation myths, and so forth.)

Why did people accept, and why do they keep accepting, these beliefs? People accept religious beliefs for many reasons. One of them is simply that once everyone in a community accepts some beliefs, voicing disagreement is often more trouble than it is worth. But there might also be a more positive reason:

because everyone not only believes in various religious entities but appears to have accepted these beliefs independently of each other, hiding the dependencies between their views and making them much more convincing in the process.

The Duna, a people several thousand strong from Papua New Guinea, believe in the kind of entities often found in traditional human cultures: ghosts and spirits that haunt the tropical forests the Duna inhabit. The Duna also share with many traditional societies origin stories about their clans.

If the Duna's beliefs are somewhat typical of those found in similar societies across the world, their language is particularly interesting in that it contains a rich system of evidentials. When they talk about ghosts and spirits, recount origin myths, or provide supernatural explanations, they must specify the source of their beliefs. And the sources they offer, as recorded by linguist Lila San Roque, are revealing.[16]

When talking about ghosts and spirits, the Duna often use evidentials denoting perception—saying that they have actually seen or heard the ghosts. The Duna have an evidential form that indicates even stronger confidence than visual perception. This evidential is used to report things you've taken part in. When you tell people, "I've had breakfast this morning," you don't leave much room for doubt. By and large, in their everyday life, the Duna use this evidential form as we would expect, to describe their past actions. But they employ the same form when relating the origin stories of their clans—which can be quite fantastical. And so they use the same marker when saying, "I've had breakfast this morning" as when they say, "Our clan's ancestors were birthed by ogres."

The use of these strong evidential forms is all the more surprising that Duna has an evidential form indicating "something people say." The Duna use this evidential when telling stories presented as fiction—the epic tales they sing for entertainment—or rumors that invite skepticism.[17] Yet, in terms of accuracy, this evidential marker would fit their beliefs about ghosts and ogres better than the evidentials they actually use.

Why do people tend to neglect the social sources of their religious beliefs? The complexity involved in acquiring religious beliefs probably plays a role. If your only source for a belief is that Amadou told you something, it is much easier to relay the source accurately than when the belief stems from repeated encounters with a group of people who share the same values and beliefs. Another factor is that the beliefs are unlikely to be challenged. Once a belief is broadly accepted within a community, there are no risks in stating it as something you're completely sure of. Maybe more important, once they have been socially acquired, religious beliefs have some (limited) cognitive, practical, or social consequences.

The epic tales sung by the Duna, and for which they use a "something people say" evidential form, are entertaining, even inspiring, but they are otherwise largely inert. By contrast, a clan's foundation myth plays a role in justifying important states of affairs, such as which land a clan claims as its own. Similarly, ghosts are used to explain misfortunes and deaths, and to justify potential retaliation.[18] Once a belief fulfills a cognitive or social role, it is easy to forget its social roots. A similar phenomenon happens when we talk about microbes or Wi-Fi signals. We have never seen either, having instead learned about them from others. Yet they are so embedded in our actions—washing our hands, picking a coffee shop with good Wi-Fi—and in our justifications for these actions, that it is easy to forget that we owe these beliefs to social sources (albeit, ultimately, scientific ones). We are more likely to tell our children, "Wash your hands, they're full of germs" than "Scientists have discovered that there are microbes that can make people sick, and that's why I believe in microbes."

While people end up accepting religious beliefs, and talking about them as if they were things they had actually witnessed or done, we shouldn't forget that not all beliefs are cognitively the same. Religious beliefs remain largely reflective rather than intuitive.[19] As you will recall, reflective beliefs only interact with a limited set of inferential or action-oriented mechanisms: they remain largely encapsulated in some part of our mind, unable to roam free like their intuitive counterparts. Otherwise, reflective beliefs would create a lot of mayhem. For instance, the fantastical origin stories about a Duna's clan interact with the mechanisms the Duna use to understand claims about whose land belongs to whom—but not with other cognitive mechanism. A Duna might believe her ancestors were ogres, but she won't prepare for the eventuality that her son might be one.

Whatever the reasons people have for professing religious beliefs as if they had acquired them on their own, how the beliefs are presented should affect their transmission. If you grow up surrounded by people who are competent at just about everything they do, are mostly benevolent, and talk confidently of having formed religious beliefs on their own, all cues should lead you to accept the beliefs. Each individual testimony would have been unconvincing (I assume you don't believe in every god of every religious person you have ever talked to), but the aggregate makes for a very persuasive package.

WHAT TO DO?

Kaluli is another of the Papua New Guinea languages with a complex evidential system. It boasts more than a dozen evidential markers, distinguishing, for instance, between firsthand, second-hand, thirdhand, and fourthhand information. The very precision of this system made it difficult to cope with new types of information sources brought by missionaries, such as book learning. Linguist Bambi Schieffelin chronicled the Kakuli's efforts to accommodate these new information sources by creating ad hoc evidentials, one of which broadly fitted *learned from a book*.[20]

The Kakuli's problem illustrates the complexity of tracing the source of a piece of information in our modern age. When we read an encyclopedia entry, what's the actual source? There is the author, but also the editor, and, more important, the numerous scholars the author relied on to write the entry, the scholars these scholars relied on, and so forth. And the problem is only getting worse—whatever the Kakuli are now doing, I hope they have a different evidential for information found on Wikipedia and information gleaned on Facebook.[21]

How can we deal with this complexity? Philosopher Gloria Origgi suggests that as "mature citizen of the digital age," we should strive to be "competent at reconstructing the reputational path of the piece of information in question, evaluating the intentions of those who circulated it, and figuring out the agendas of those authorities that leant it credibility." We should adopt a reflective attitude toward sourcing, much like professionals, asking ourselves of a new piece of information "Where does it come from? Does the source have a good reputation? Who are the authorities who believe it? What are my reasons for deferring to these authorities?"[22] Doing this work would also help us uncover hidden dependencies: as we track the provenance of different opinions, we are in a better position to realize when they all stem from the same source.

Besides doing this detective work, we should also help others by providing accurate sources for our opinions. It is tempting to try to get as much credit for our opinions as possible. When we possess a relevant piece of information—some political news, a scientific fact, a pertinent statistic—or even when we convey our opinion on complex issues, we should attempt to provide sources as accurately as possible. This would often mean minimizing our own role in the process.[23]

Being more open about how we formed our beliefs not only would help others decide for themselves whether or not they should believe us but also would help them better understand which sources are reliable. My guess is that if we were more accurate in reporting our sources, Wikipedia, the "mainstream media," and other sources that are sometimes scorned would get much more credit than they currently do, while each of us would get a little less—and rightfully so. In turn, even if we grant less credit to people who disclose their sources—since the ideas aren't their own—we should still be thankful to them, as they improve our informational environment.

12

WITCHES' CONFESSIONS AND OTHER USEFUL ABSURDITIES

ON NOVEMBER 17, 1989, the body of fifteen-year-old Angela Correa was found in a park in Peekskill, upstate New York. She had been raped, beaten, and strangled. Jeffrey Deskovic, a fellow student of seventeen, reacted very emotionally to her death, attracting the investigators' attention. Brought in for questioning, he ended up confessing to the crime.

More than a year later, Deskovic's trial was ending. Material evidence suggested he hadn't done it—in particular, his DNA did not match that of the semen found in the victim's body. But he had confessed, and that was enough to sway the jury, which found him guilty. He was sentenced to fifteen years to life.[1]

Even though Deskovic had retracted his confession, the district attorney (now Fox News host Judge Jeanine) refused to authorize more DNA testing that might have pointed to another culprit and exonerated Deskovic. It was only when a new DA took office, in 2006, that more DNA tests were run, suggesting that Steven Cunningham, who was already serving a prison sentence for murder, had raped and killed Angela Correa. Cunningham was convicted, Deskovic exonerated. He had spent sixteen years in jail.

Estimating the rate of false confessions is difficult: unless a case is overturned, it is hard to tell whether a confession was true or false.[2] Some estimates point to a low number for minor offenses: a few percent.[3] Others suggest a higher proportion for more severe offenses: more than 10 percent for people currently imprisoned.[4] What we know for sure is that false confessions are distressingly common among people who have been later exonerated: between 15 and 25 percent, and even more for the most serious crimes such as homicide.[5]

Confessions, true or false, are incredibly persuasive. They are more convincing than the most influential other form of evidence: eyewitness testimony.[6] Confessions are so definitive that even when they are retracted, the accused is overwhelmingly likely to be convicted.[7] As legal scholar Charles McCormick put it in his *Handbook of the Law of Evidence*, "The introduction of a confession makes the other aspects of a trial in court superfluous."[8]

Ultimately, the relevant question here is: Why do our mechanisms of open vigilance seem to be blind to the possibility of false confession? But I must first address what seems to be an even more puzzling question: Why do people confess to crimes they haven't committed?

False confessions are offered for a variety of reasons. Many voluntary false confessions are given to cover for someone else.[9] Some have weirder motives, from impressing one's partner to hiding an affair.[10] But most false confessions aren't fully voluntary. Instead, they are coerced through a variety of means, ranging from old-school physical abuse, to raising expectations of leniency, or even promising small, immediate rewards, "being allowed to sleep, eat, make a phone call, go home."[11] In some jurisdictions—such as some U.S. states—the interrogators can also lie to suspects, telling them that the police hold overwhelming evidence against them. By skillfully highlighting the short-term gains, while lessening the long-term costs, investigators can make it seem like a confession is a suspect's best option, even if they haven't done anything wrong. When Jeffrey Deskovic confessed, he had been, by his account, yelled at for hours, threatened with the death penalty, and told that if he confessed the abuse would stop and he would be sent to a mental hospital.[12] That was enough to break an emotionally fragile teenager (indeed, false confessions overwhelmingly come from young and/or mentally challenged people).[13]

Whatever the reasons, false confessions are routinely produced, and are then generally believed. For our mechanisms of open vigilance, a confession ticks all the boxes for a reliable message. First, speakers are supposed to be competent, as they simply have to report on things they have done themselves. We might understand that someone doesn't remember doing something they have in fact done, but to falsely remember doing something (often horrible) that they haven't done stretches the imagination. Second, speakers are expected to be honest. Our sensitivity to speakers' incentives means that self-interested statements—such as denials of wrongdoing—are sure to be heavily discounted, but also that self-incriminating statements are easily accepted.

To make things worse, people are just as incapable of discriminating between true and false confessions, on the basis of the accused's demeanor, as they are between truths and lies more generally (for reasons explained in chapter 6). The police and laypeople perform equally poorly in this respect, the only difference being that professional interrogators are much more confident in their abilities, even though this confidence is unjustified.[14]

By default, it makes sense that confessions should be believed. But why aren't they discounted when made under pressure? As a matter of fact, they are, when the pressures are strong and transparent. Psychologists Saul Kassin and Lawrence Wrightsman presented participants with the transcript of a trial in which the accusation mostly rested on the defendant's confession.[15] When the confession was obtained after a threat—that the defendant, if he didn't confess, would be treated poorly and get a maximum sentence—it was essentially ignored by the participants.

Unfortunately, the people who ultimately have to judge the convincingness of a confession often do not have access to all the relevant information.[16] It used to be relatively easy, and it still is in many jurisdictions, for interrogators to pressure suspects without the judge or the jurors knowing about it. It is also difficult for judges and jurors to fully grasp the suspect's emotional state, the strain exerted by hours of close interrogation, or the yearning for any slight reprieve. The relative paucity of information about the pressures bearing on suspects means that judges and jurors easily revert to the default stance of accepting self-incriminating statements.

Yet most, if not all, of these exculpatory elements are known to the interrogators who have obtained the confessions, and that doesn't stop them from accepting the confessions. As a rule, when interrogators seek to obtain a confession, their prior inquiries have led them to believe the suspect is guilty. At this stage the interrogators' goal, as they see it, might be less to ascertain guilt than to build a convincing case. As a result, they are likely to use their mechanisms of open vigilance to gauge their chances that the confession will be accepted by judge and jury, rather than to evaluate it critically. Given that judge and jury likely won't be aware of many of the elements that would make the confession less persuasive, it makes sense that the interrogators deem the confession acceptable as well. Moreover, less obvious pressures—raising the suspect's expectation of leniency, say—aren't considered by juries even when they are known: on their own, such pressures don't appear sufficient to explain why someone would admit to a heinous act and risk years in jail.[17] As long as the most egregious violations of the suspect's rights aren't recorded, it is relatively easy for interrogators to provide convincing confessions.

If confessions (true and false) already play an important role in securing criminal convictions in the United States, in countries with more lax interrogation standards, or stronger social pressures, confessions can prop up the entire system of criminal law. In Japan, more than 99 percent of cases sent to trial in a year can end up in convictions, with around 90 percent of convictions based on confessions.[18] Although it is impossible to tell how many of these confessions are false, some particularly egregious cases are well known. An investigation in the late 1970s targeted thirty-six minors accused of leading a violent biker gang. By the end of the interrogations, thirty-one of them had confessed to being one of the three gang leaders.[19] Still, if Japanese suspects sometimes plead guilty to breaking laws they haven't broken, at least they are human laws, not the laws of physics—which witches routinely confess to transgressing.

EXTRAORDINARY CONFESSIONS AND THE MADNESS OF WITCHES

The Salem witch trials, related in Arthur Miller's famous play *The Crucible*, started with Tituba, an enslaved woman who had been brought from the Barbados to Salem in 1680, and who was accused of bewitching two young girls. She barely bothered denying it. Soon confessions poured out of her like fantastic curses. She had been "rid[ing] upon a stick."[20] Sarah Osborne, one of her accomplices, had a creature with "wings and two legs and a head like a woman," as well as a kind of werewolf ("a thing all over hairy, all the face hairy, and a long nose, . . . and goes upright like a man").[21] Tituba would end up implicating hundreds of people, fueling the most famous witch hunt in American history.

As we can safely assume Tituba's confession to be false, we could ask the same questions as we did previously: Why did she confess? And why was she believed? However, given the extraordinary nature of the charges, we must start by wondering why she was accused at all.

Although witchcraft beliefs vary from one culture to the next, the core concept that some people are able to hurt others through supernatural means is extraordinarily common across a wide range of societies. As anthropologist E. E. Evans-Pritchard noted in his landmark study of witchcraft among the Azande (of Central Africa), a belief in witchcraft does not replace a more commonsensical understanding of causation. In Evans-Pritchard's classic example, a Zande would know that a hut fell down because time and termites had gnawed at its pillars. They would also know that the hut fell down at such and such a time, injuring such and such individuals who happened to be sitting under it, because of a witch (I will use *witch* here to refer to either gender, as is usually done in the anthropological literature).[22]

Why add this layer of intentionality to random events that are well captured by other intuitions? It makes sense for our minds to overinterpret misfortunes in intentional terms. Better to look for a culprit when there was no foul play than to let someone hurt us without suspecting anything. After all, some people do bear us ill will, and we often know it. As a result, when something bad happens to us, and even more so when we suffer a succession of misfortunes, it seems appropriate to look for an agent that might have caused them, our enemies being prime suspects.

Imagine you work in an office and have a serious grudge with Aleksander, one of your colleagues. You have already had minor skirmishes, playing nasty pranks on each other. One day, you can't find your stapler, the drink you had put in the fridge has a weird taste, and your computer keeps crashing. Wouldn't the thought that your office nemesis is responsible pop into your mind? Still, there's a world between these plausible—if a bit paranoid—suspicions and a belief that one is being poisoned by stick-riding witches with werewolf sidekicks. Confessions might have helped bridge this gap.

Imagine you've become quite persuaded Aleksander is responsible for your office troubles. If he doesn't care one bit about how you perceive him, he will deny (truthfully, let's say) any involvement. But if at some point he needs to make peace, his best option might be to get you a new stapler, buy you a drink, and fix your computer—which would be taken as a confession of guilt, whether he explicitly admits to the misdeeds or not.[23] Only then could you start to forgive him and move on.

Now, because you have some basic knowledge of medicine if, on top of your other misfortunes, you had caught the flu, you wouldn't blame Aleksander for it. But if you did not have that knowledge, you might have lumped this extra misfortune with the others. If Aleksander had also confessed to making you sick, you would then have been tempted to form a belief that some people can make others sick at will.

This tentative sketch shows how a belief in something like witchcraft could emerge from a cycle of suspicion, the need to mend fences, and false confessions.[24] Still, confessing to stealing an officemate's stapler is one thing, but witches are routinely accused of vastly graver offenses, even murder. Why would anyone confess if the penalty is, say, being burned at the stake? If it is true that this gruesome penalty was common in early modern

Europe, in many societies witches who confess receive rather lenient sentences.

Among the Azande, presumed witches are made aware of the accusations against them when a fowl's wing is placed in front of their house. The witch is then expected to take the wing to the victim's house, blow some water on it, confess to the bewitching, and apologize.[25] The Ashanti (Ghana) ask the witch to make a public confession and pay a fine.[26] A Banyang (Cameroon) suspected of being a witch is made to dance at a specific rhythm.[27] Among the Tangu (New Guinea), the witch must compensate the victim.[28]

There are multiple reasons for keeping the penalties inflicted on witches low. Harsh penalties are difficult to implement and might lead to retaliation by the accused witches and their allies. Confessing not only is often low cost but can also provide some advantages. By confessing, a witch can "win mercy and forgiveness."[29] Since many accused witches confess, even without threat of duress, it is quite plausible that "confessions are crucial" for belief in witchcraft to flourish, as anthropologist Roy Willis argued.[30]

Once a belief in witchcraft has become ingrained in a culture, it becomes possible to justify punishing a suspected witch even in the absence of a confession, making confessions even more worthwhile—after all, the whole point of confessing is to be treated better, given that others are persuaded we are guilty.[31]

A Zuni (Native American) boy was accused of bewitching a young girl who had had a seizure after he had touched her hands. He knew witchcraft called for the death penalty.[32] Just like Tituba's, his initial denials proved unconvincing—he would say he was innocent, wouldn't he? And so, instead, the boy made up a story of having been taught witchcraft, and he attempted to cure the girl. As the trial dragged on, he concocted increasingly baroque tales of taking on animal forms and killing people by spitting cactus needles at them. He ended his confession by bemoaning the recent loss of all his powers. Whether they were amused, frightened, or impressed by his frankness, the judges freed him.

Returning to Tituba, she remained in jail for a few months but was eventually freed without even being indicted. Indeed, none of the women who confessed were among the nineteen who hanged in Salem. Even at the height of the witch hunt in England, the choice for a witch would often either be "to confess her guilt and promise amendment of life at the ecclesiastical courts, or [be] removed from the community by imprisonment or death at the Assizes."[33]

Given the crimes witches stood accused of, the most remarkable aspect of their punishment is its leniency—provided the witches confess and make amends. Witches were often let go with a confession, the odd ritual, maybe a small fine, even though their avowed crimes included making people sick, killing them, poisoning their crops and cattle, conspiring with the devil, or even devouring their own children.[34] Indeed, in some cultures, such as the Azande, after the confession had been made the witch could fully reintegrate into society as if nothing had happened.[35] Presumably, few people would treat someone in this way whom they actually saw poisoning their food or eating children.

In this respect, beliefs in witchcraft behave like the wild rumors discussed in chapter 10. People do not draw from these accusations all the conclusions they would if the accusations were intuitively believed, for instance, because they were based on perception. Beliefs in witchcraft remain reflective, not fully integrated with the rest of cognition. When witches are executed, the accusations of witchcraft aren't the only, or even arguably the main, driver, playing more of a post hoc justificatory role. Instead, the usual self-interested motives rear their ugly heads.

For example, in Tanzania, witch killings increase in times of drought or flooding, and they mostly target elderly women perceived as a weight on the family.[36]

HOW TO BE A CREDIBLE SYCOPHANT

Confessions, even the most extraordinary ones, are convincing because they are self-incriminating. In a more roundabout manner, the intrinsic believability of self-incriminating statements also explains why people profess the silliest opinions, from lavishing absurd praises on Kim Jong-il to claiming (nowadays) that the earth is flat.

It has been told of Kim Jong-il, father of current North Korean leader Kim Jong-un, that he could already walk and talk when he was a tiny babe of six months.[37] At university, he wrote more than a thousand books and articles. His perfect memory allows him to remember "all the exploits performed by the famous men of all ages and countries, all the political events, big and small, and the significant creations of humankind and their detailed figures, [as well as] the names, ages, and birthdays of all the people he has met." He understands all complex topics "better than the experts."[38] Kim Jong-il can also teleport, control the weather, and set worldwide fashion trends.[39]

In terms of overinflated compliments, Kim Jong-il is hardly unique. According to the most inventive toadies of their respective countries, Hafiz al-Assad (Bashar's father) was Syria's "premier pharmacist"; Nicolae Ceaușescu was the Giant of the Carpathians, the Source of Our Light, the Celestial Body; Mao could easily beat swimming world records; Saddam Hussein was the new Nebuchadnezzar.[40]

Could people really have been brainwashed into believing such nonsense? Obviously not. Even in North Korea, "few people have been convinced by this propaganda because since Kim [Jong-il] came to power, economic conditions have gone from bad to worse."[41] The Romanians who had been heaping praises on Ceaușescu were all too eager to lynch the Source of Their Light™ when the opportunity arose. Libya's citizens might have plastered Gaddafi's face on their walls, their halls, their stalls, they still hunted him like a wild animal when his regime crumbled. Flattery toward these Dear Leaders was not "a way of expressing deeply held emotions," but "a code to be mastered" if one wanted to survive in ruthless regimes.[42]

Leaders don't incentivize such flattery because they expect people to believe the lavish praises. Indeed, the leaders don't believe the praise themselves: as Mao advised Ho Chi Minh, "The more they praise you, the less you can trust them."[43] There are, however, some exceptions, cases in which shows of support and hyperbolic flattery can be reliable signals of commitment: when they are so over the top that they help the speaker burn their bridges with other groups, thereby credibly signaling their allegiance to the remaining group.

BURNING BRIDGES

Joining a group, which could be anything from an amateur soccer team to a clique at work, brings benefits: being supported and protected by other members, performing activities that wouldn't be possible on your own. These benefits come with a cost: a member must do their bit to support the other members and contribute to joint activities. A member of a soccer team is expected to show up to training sessions, give their best during matches, and so forth.[44]

Because being part of a good group has many advantages, its members have to exert caution toward new potential recruits, to make sure the recruits are willing to pay the costs and not simply ready to reap the benefits.[45] The soccer team doesn't want people who are only interested in playing the odd match when they feel like it.

When someone wants to join a group, it can be difficult to honestly signal their willingness to pay the costs associated with membership. They can say, "I'll be a good team player!," but that's unlikely to be very persuasive. Such a statement is only credible to the extent that the speaker's commitment is taken seriously. In turn, respect for the speaker's commitment hinges on whether we expect them to be good member of the group. In other words, if we think the speaker will be a good group member, we believe them when they say so, but if we don't think they will, we don't believe them. As a result, the statement is useless.

There are many ways for a new recruit to demonstrate their commitment to being a good group member. For instance, they can endure an initial phase in which the costs are higher than the benefits—attending training sessions but remaining on the bench during matches, say. Another solution is to signal disinterest in the alternatives by burning their bridges. If you are a gifted amateur soccer player who can take your pick from many soccer teams, the members of any one team might doubt your loyalty: you could easily change your mind and join one of the other clubs. If, however, you are really motivated to join a particular club, you could prove your loyalty by publicly disparaging the other clubs.

The statement "I really don't want to join your group," made to members of that group, is quite believable. It is another kind of self-incriminating statement: Why would someone say something like this if it were not true? Such statements can easily become more credible by being insulting. Someone who says, to their face, "I hate your group and everything it stands for" is, not surprisingly, unlikely to ever be accepted by the members of that group. By burning your bridges with as many of the competing groups as possible—making you unclubbable, as cognitive scientist Pascal Boyer put it—you credibly signal to the remaining groups that you'll be loyal to them, since you don't have any other options.[46]

Some extreme flattery likely stems from the application of a burning bridges strategy. When a writer suggests that Kim Jong-il can teleport, he doesn't expect his audience (least of all Kim Jong-il) to literally believe that. The point, rather, is to make the groveling so abject that even other North Koreans find it over the top. By signaling to other North Koreans that he's willing to go beyond what's expected in terms of ridiculous praises, the writer is telling the audience that he would rather seek Kim Jong-il's approval than that of a broader base of more sensible people, who only say Kim Jong-il can influence the weather, but not teleport. As a result, the writer is credibly signaling his loyalty to Kim Jong-il.

Over-the-top flattery is far from the only way of making oneself unclubbable. Other statements that make one look incompetent to everyone except a select group can be used. A philosopher from Cardiff University recently claimed that evolutionary biology and genetics were just as (un)scientific as creationism.[47] A scholar at Scripps College in the United States argued against the "human/non-human binary that undergirds ... biological conceptions of life" and suggested pandemics weren't due to the usual suspects (such as "poor hygiene") but instead were the result of "global industrial resource extraction."[48]

These views are roundly rejected by the relevant experts in each domain—indeed, they are rejected by the vast majority of scholars. As a result, stating them makes one unappealing to most academic departments. By adopting these positions outside the norm of what is accepted in academia, however, these intellectuals may have sought to enhance their positions within a network of postmodern scholars, who tend to hold relativistic views about the truth, and who are often opposed by the rest of the scientific community. In a different vein, people from modern societies who proudly proclaim the earth is flat are pretty sure to be ridiculed by most, but also to be seen as loyal members of the small (but growing!) flat-earther community.

Making statements that the majority find morally repellent is also a good way of burning bridges. Many are offended by extreme libertarian views, such as saying that taxation is slavery, or arguing, as does economist Murray Rothbard, that laws shouldn't punish parents who starve their children to death.[49] Others are shocked by the pronouncements of Holocaust deniers.[50] Vast audiences have been scandalized by the threats of ISIS recruits, such as that proffered by a British convert: "When we descend on the streets of London, Paris and Washington the taste will be far bitterer, because not only will we spill your blood, but we will also demolish your statues, erase your history and, most painfully, convert your children who will then go on to champion our name and curse their forefathers."[51]

How do we know that these extreme positions—from stating the earth is flat to denying the Holocaust—are a way of burning bridges? Couldn't they instead stem from a process of personal inference (people see the horizon as flat; they can't imagine that something like the Holocaust could happen) or of persuasion (seeing YouTube videos defending flat-earth theories, reading a book by a Holocaust denier)?

A first argument in favor of the burning-bridges account is the sheer extremity of the views being defended. We're dealing with positions the vast majority of the population finds either blatantly stupid or irredeemably evil. Still, some scientific positions might have been perceived in the same way: most people find intuitively ludicrous the idea that humans are descended from fish, say. But the burning-bridges strategy adds insult to injury by impugning the intelligence or moral standing of those who disagree with the beliefs used to burn bridges. Extreme postmodern thinkers not only appear a little crazy to most but also suggest that those who fail to agree with their arguments are unsophisticated fools. Holocaust deniers make morally repellent claims but also paint those who disagree as enraged Zionists or their useful idiots. Holding such positions is a surefire way of making oneself unclubbable by all but the small clique that defends similar views.

Still, even in the burning-bridges account, it is not obvious why any group would find such views palatable in the first place. For beliefs to work in the burning-bridges scenario, they have to be extreme. This creates an incentive for new recruits, or even for members who wish to improve their status in the group, to push the limit of what the group already finds acceptable. The positions just mentioned are so extreme because they are the outcome of a runaway process in which increasingly bizarre views must be defended. When Kim Jong-il was starting to consolidate his grip on power, someone who claimed he could teleport would have been seen as cuckoo. It's only after many rounds of flattery inflation have led to a group of people who agree that Kim Jong-il can control the weather that claiming he can teleport makes some kind of sense (I owe the term *flattery inflation* to Xavier Márquez).[52] The same goes for all the other positions. No one jumped from "We might want to rethink the legitimacy of some legal constraints" to "For instance, why does the law punish parents who starve their children?" or from "Scientific progress is more complex than the typical Whiggish history allows" to "And so everything is relative and there is no truth." In each case, there were many steps before these heights of inanity were reached, each one making steadily more extreme views more acceptable.[53]

It is difficult to believe that people would publicly and confidently profess absurd or repugnant views. But stating our views publicly and confidently is precisely what is required to become unclubbable. The groups we want to burn bridges with must know we hold unpopular or offensive views, and the groups we want to join must know that the other groups know. Being a closet flat-earther isn't going to give anyone the keys to the flat-earther's country club. By contrast, if people came to hold these extreme views through other means—personal inference or persuasion—they would realize going public might reflect poorly on them, and would be more discreet.

Finally, as was the case with most beliefs discussed in the last two chapters, burning-bridges beliefs are held reflectively. The person who said Kim Jong-il could teleport would presumably be very surprised if Kim beamed up in front of him, *Star Trek*–style. Postmodern thinkers who believe all truth to be relative still look at a train timetable before going to the station. People who hold such beliefs are very vocal, and appear very confident, not because these are intuitively held beliefs—beliefs they would let freely guide their inferences and decisions—but because that's how you burn bridges.

Defending extreme beliefs as a way of burning bridges isn't a failure of open vigilance, as it would be if the defenders of these beliefs had been talked into intuitively accepting them. Instead, it reflects a perverse use of open vigilance. We can use our open vigilance mechanisms to anticipate what messages others will likely accept. As a rule, if we anticipate rejection, we think twice before saying something. When we want to burn bridges, we do the opposite: the more rejection we anticipate—from all but the group we would like to join—the more likely we are to voice our views. This perverse use of open vigilance mechanisms doesn't have to be conscious—indeed, in the vast majority of cases I imagine it isn't. It appears to be quite effective nonetheless.

WHAT TO DO?

Self-incriminating statements are intrinsically credible. Because they refer to our own beliefs or actions, we're supposed to know what we're talking about. Because they make us look bad, we would have no reason to lie.

If believing self-incriminating statements is, on the whole, a good heuristic, it also leads to a series of problems. The most obvious are the false confessions that plague judicial systems. The answers here are mostly institutional: the law should reduce as much as possible the pressures put on suspects, and make whatever pressures are left as transparent as possible for judges and jurors to consider. For example, in the United Kingdom, it is illegal for the police to lie to suspects, the whole interrogation has to be taped, and dubious confessions are likely to be suppressed before they reach the jury.[54]

More generally, we should keep in mind that people might confess to regain our approval, even if they haven't done anything wrong. In such cases, we should believe in the social goals (they are willing to make peace with us) rather than the content (they have really done the thing they confess to). In the end, it is these goals that matter the most.

The same logic applies to the self-incriminating statements that are used for burning bridges. We shouldn't assume that people intuitively hold the apparently deranged or evil views they profess. However, we should take seriously their social goal, namely, to reject the standard groups that make up the majority of society in favor of a fringe coalition. As a result, if we want them to abandon their silly or offensive views, attempting to convince them of these views' logical, empirical, or moral failings is unlikely to work. Instead, we have to consider how to deal with people who feel their best chance of thriving is to integrate into groups that have been rejected by most of society.

People aren't stupid. As a rule, they avoid making self-incriminating statements for no reason. These statements serve a purpose, be it to redeem oneself or, on the contrary, to antagonize as many people as possible. By considering the function of self-incriminating statements, we can react to them more appropriately.

13

FUTILE FAKE NEWS

GALEN, WHO WENT FROM treating wounded gladiators to serving Roman emperors, was undoubtedly a brilliant physician and a skilled surgeon. His dissections (and vivisections) advanced our understanding of anatomy, and his ideas affected Arabic and Western medical thought for more than a thousand years. But Galen was also a staunch supporter of the humoral theory of disease.[1] This theory explains mental and physical diseases as resulting from an imbalance between the four humors contained in our bodies: blood, yellow bile, black bile, and phlegm. Blood, because it was thought to contain elements of the other three humors, was considered to be the best leverage to restore balance between them, and therefore bring back health.[2] Since transfusion wasn't a practical option, bloodletting—cutting open a vein to let blood flow out—was commonly used to remove the excess humor. In line with the humoral theory, Galen was rather generous with his bloodletting prescriptions, recommending this therapy for gout, arthritis, pleurisy (inflammation of the tissues around the lungs), epilepsy, apoplexy, labored breathing, loss of speech, phrenitis (inflammation of the brain), lethargy, tremor, depression, coughing blood, and headache. He even recommended bleeding as a cure for hemorrhages.[3] Galen's defense of the humoral theory proved widely popular, dominating Western medicine from the eleventh century, when his texts found their way to the nascent European universities, up to the nineteenth century, when the theory was finally debunked.

Looking back at the accusations against Jewish shopkeepers that flourished in Orléans in the spring of 1969, we are tempted to make fun of the people who believed such tall tales. Local shopkeepers sending young girls to be prostituted in faraway countries? Please! After all, the rumor didn't really hurt anyone. Before Easter 1903, in Kishinev (currently Chișinău, capital of Moldova), accusations circulated about the local Jewish population, rumors that the Jews had murdered a child and drained him of his blood in a religious ritual.[4] If the rumors of blood libel were as ludicrous as those of Orléans, they appeared more consequential. The inhabitants of Kishinev didn't just gossip and glare at the suspects. They struck ferociously, killing scores in the most gruesome fashion, raping dozens of women, pillaging hundreds of stores and houses. The world over, rumors of atrocities, such as the blood libel, are a prelude to ethnic attacks.[5]

In 2017, the *Collins* dictionary designated *fake news*, information that has no basis in fact but is presented as factual, its word of the year.[6] This decision was a reaction to the abuse of fake news in two events that took place in 2016: the election of Donald Trump to the U.S. presidency, and the decision made in the United Kingdom, by referendum, to leave the European Union (Brexit). In both countries, a large majority of the elites and the traditional media, surprised and dismayed by people's choices, searched for explanations. Fake news was a common answer. "Fake News Handed Brexiteers the Referendum" was the title of an article in the *Independent*, a British newspaper. Across the Atlantic, the *Washington Post* ran a piece claiming, "Fake News Might Have Won Donald Trump the 2016 Election."[7] Even when it is not about politics, fake news is scary: a piece in *Nature* (one of the world's foremost scientific publications) suggested that "the biggest pandemic risk" was "viral misinformation."[8]

Some fake news spread the old-school way, carried, for instance, by "Brexit buses" claiming the United Kingdom was sending £350 million a week to Brussels that could be redirected to the health services instead (in fact, the number is nowhere near that high, and most of the money goes back to the United Kingdom anyway).[9] But fake news, which has always existed in one form or another, was seen as particularly threatening this time around because social media had vastly expanded its reach.[10] In the three months leading up to Donald Trump's election, the twenty most popular fake news stories related to the election garnered more than eight million shares, comments, and likes on Facebook.[11] Among the most popular fake news were stories about Hillary Clinton, Trump's opponent, selling weapons to the terrorists of ISIS, or the pope endorsing Trump. Through the sharing of fake news, and of partisan news more generally, social media have been accused of creating echo chambers that amplify people's prejudices and polarize the population, leading to extreme political views.[12]

What do the humoral theory of disease, blood libels, and Trump's endorsement by Pope Francis have in common? Obviously, they are inaccurate pieces of information. They are also linked with outcomes ranging from the clearly terrible (ethnic attacks, the systematic mistreatment of patients) to the arguably suboptimal (Trump's election, Brexit). It would be natural to think that these false beliefs led directly to the outcomes described: physicians practice bloodletting because they accept the humoral theory of disease; ethnic minorities are massacred because of the atrocities they are accused of committing; people vote the "wrong" way because they are misled by fake news.

If this were the case, we would be dealing with very grave failures of our open vigilance mechanisms, in which people would have been persuaded to accept the misleading ideas of influential physicians, rumormongers, and fake news purveyors. Unlike some of the beliefs described in the previous chapters, these misleading ideas would have dramatic consequences not only for others but also for those who hold them: physicians who ask to be bled, perpetrators of ethnic violence who get hurt, and people who end up voting against their interests.

In this chapter, I argue that this account gets the direction of causality wrong. By and large, it is not because the population hold false beliefs that they make misguided or evil decisions, but because the population seek to justify making misguided or evil decisions that they hold false beliefs. If Voltaire is often paraphrased as saying, "Those who can make you believe absurdities can make you commit atrocities," this is in fact rarely true.[13] As a rule, it is wanting to commit atrocities that makes you believe absurdities.

EVERYBODY BLEEDS

Reading David Wootton's *Bad Medicine* was an eye-opening experience, revealing how until around a century ago doctors were not only useless but positively harmful, and arousing my interest in bloodletting.[14] How could this practice have been accepted for so long? My initial reaction was to trace it back through the great physicians who had defended it, from Benjamin Rush in nineteenth-century America to the Hippocratic writers in ancient Greece. A specific link in this chain was fascinating: from the eleventh century onward, hundreds of thousands of people would be bled because a couple of Galenic manuscripts on the humoral theory of disease survived down the centuries to reach the first European medical schools.

But as I started looking into the anthropological literature, I was quickly disabused of my Western-centric views. Far from being a historical anomaly, bloodletting was practiced all around the globe, by people who had never heard of Rush, Galen, or the Hippocratic writers. The Guna (Panama and Colombia) used a miniature bow to shoot a miniature arrow into the temple of those suffering from headaches. When someone complained of headaches, abscesses, or chest pain, the Bagisu (Uganda) sucked a bit of blood from the ailing area with a hollowed horn. The Iban (Malaysia) cut a small incision in the back when someone was afflicted with back pain. The Dayak (Borneo) relied on a heated bamboo to draw blood from any ailing part of the body. Bloodletting was also practiced by major non-Western civilizations, playing a role in ancient Indian and Chinese medicine.[15]

All in all, at least a quarter of the world's cultures likely practiced some form of bloodletting at some point in their history. In some of them—ancient Greece, ancient China—the practice was accompanied by complex theoretical explanations. In most, however, people were content with a cursory "We've got to let the bad stuff out."[16] If the humoral theory of disease can't explain why bloodletting spread in 99 percent of cultures, which have never heard of humors, it does not explain either why it spread in the cultures that embraced the humoral theory. Galen developed sophisticated theories to justify something people wanted to do anyway: when they are sick or in pain, let a bit of blood flow to evacuate hypothetical internal pollutants.

If bloodletting can be found throughout the world without its Western trapping, the humoral theory, by contrast, rumors of atrocities seem to be a standard component of the ethnic riot, suggesting that these rumors play a significant causal role.[17] In fact, the arrow of causality is unlikely to point in this direction, as there is little fit between the rumors and the violence. We find countless instances of rumors not followed by any violence, and when violence does happen, its nature is typically unrelated in form or degree to the content of the rumors.

When the Jewish population of Kishinev was accused of the murder of a small boy, the lie took hold because people broadly believed this ritual to be "part and parcel of Jewish practice."[18] Indeed, alarming rumors surfaced every year before Easter, without any attendant pogrom.[19] Shouldn't this strike us as bizarre? Who harbors in their midst people suspected of periodically kidnapping children to bleed them to death? That the same beliefs did not lead to violence most of the time suggests the beliefs themselves do not explain why the violence erupted when it did.

If the local Christian population in Kishinev had genuinely believed in the blood libel, we might have expected some terrible reprisal, maybe the murdering of Jewish children, or of the adults thought to be guilty. The reprisal is terrible indeed, but it bears no relation to the accusations: How is pillaging liquor stores going to avenge the dead child? In other times and places, Jewish populations have been massacred, women molested, wealth plundered under vastly flimsier pretexts, such as accusations of desecrating the host. Even in Kishinev, the allegations piled up with no sense of proportionality, from killing children to dishonest business practices: "Those awful Jews. They bleed our children to death. And they cheat us on the change!" By and large, scholars of rumors and of ethnic riots concur that "participants in a crowd seek justifications for a course of action that is already under way; rumors often provide the 'facts' that sanction what they want to do anyway."[20]

What about fake news, then? Can it sway momentous political decisions? Here I focus on the election of Donald Trump, the event for which the most data are available. At the individual level, there was a correlation between viewing fake news websites, which overwhelmingly supported Trump, and being a Trump supporter.[21] At the state level, the more people visited fake news websites, the more likely the state was to vote for Trump.[22] Does this mean that viewing fake news prompted people to vote for Trump? Not necessarily. The majority of people who visited fake news websites weren't casual Republicans but "intense partisans," "the 10% of people with the most conservative online information diets."[23] These people were very unlikely to have turned from Hillary voters to Trump supporters. Instead, they were scouting the web—not only fake news websites but also the traditional press—for ways of justifying their upcoming decision to vote for Trump, or of demonstrating their support.[24]

A study by Brendan Nyhan and his colleagues supports this interpretation.[25] Trump supporters were provided with accurate information correcting some of Trump's false statements (rather than fake news, but the principle is the same). Most of them accepted the corrections. Yet the supporters didn't waver in their support for Trump. This suggests that the initial acceptance of the false statements hadn't caused their support for Trump. Rather, they had accepted the statements because they supported Trump.

Political scientists Jin Woo Kim and Eunji Kim observed a similar pattern when they studied the rumors that Barack Obama is a Muslim, which circulated prior to the 2008 presidential election pitting Obama against John McCain.[26] Kim and Kim compared the answers to two waves of political surveys: one taken before the rumors started to spread, and one after they had peaked. The researchers found that the rumors did have an effect: they made people more likely to say Obama is a Muslim.

However, this was only true for people who were already inclined to dislike Obama. As a result, the rumors had no effect on people's general attitude toward Obama, or on the likelihood of voting for him: accepting the rumor didn't make people dislike Obama; disliking Obama made people accept the rumor.

A REASON FOR EVERYTHING

If people are going to do whatever they want anyway—from practicing bloodletting to attacking their neighbors—why would they bother with a variety of absurd and inert beliefs? Humans are an uber-social species, constantly evaluating each other to figure out who would make the best cooperation partners: who is competent, who is nice, who is reliable. As a result, we're keen to look our best, at least to people whose opinions we value. Unfortunately, we're bound to do things that look stupid or morally dubious. When this happens, we attempt to justify our actions and explain why they weren't, in fact, stupid or morally dubious. This lets us correct negative judgments, and it helps observers better understand our motives, thus judging us more accurately.

We not only spontaneously justify ourselves when our behavior is questioned but also learn to anticipate when justifications might be needed, before we have to actually offer them.[27] This creates a market for justifications. But such a market arises only when we anticipate that some decisions are likely to be perceived as problematic.

As mentioned previously, the small-scale societies that practice bloodletting typically do not elaborate complex theories to justify the practice; it is simply seen as the obvious option when someone is suffering from a particular ailment. By contrast, in larger or more diverse communities, alternative treatments are sure to be in competition, and physicians as well as patients have an incentive to justify their decisions. This competition and the attendant debates were certainly important in ancient Greece, where the humoral theory of disease was developed by the Hippocratic writers.[28] The same competition existed in Rome when Galen set up shop: it was only after his treatments were questioned by the local doctors that Galen developed a book-length defense of bloodletting, drawing on his Hippocratic predecessors.[29] In a small-scale society, you can practice bloodletting no questions asked, but in more sophisticated cultures, to go around bleeding sick people, you need a theory.

As for fake news, it also flourishes, when needed, as a form of justification.[30] In 2016, the year of the presidential election, six of the top ten most shared fake news stories on Facebook were political, from the pope's endorsement of Trump to an ISIS leader's endorsement of Clinton.[31] By contrast, in 2017, only two of the top ten fake news stories were political (including a rather funny "Female Legislators Unveil 'Male Ejaculation Bill' Forbidding the Disposal of Unused Semen").[32] Furthermore, more than 80 percent of fake news related to the 2016 elections was pro-Trump, and conservatives were more likely to share fake news on social media.[33] The abundance of pro-Trump fake news is explained by the dearth of pro-Trump material to be found in the traditional media: not a single major newspaper endorsed his candidacy (although there was plenty of material critical of Clinton as well). At this point, I should stress that the extent to which fake news is shared is commonly exaggerated: during the 2016 election campaign, fewer than one in ten Facebook users shared fake news, and 0.1 percent of Twitter users were responsible for sharing 80 percent of the fake news found on that platform.[34]

Some political fake news—for instance, "WikiLeaks: Clinton Bribed 6 Republicans to 'Destroy Trump'"—might sound plausible enough, at least to people with little knowledge of politics; that is, most of the electorate. But many stories would presumably sound quite absurd to almost everybody (e.g., "[Evangelical leader Franklin] Graham Says Christians Must Support Trump or Face Death Camps"). In this respect, political fake news resembles other fake news. In 2017, the biggest hit was "Babysitter Transported to Hospital after Inserting a Baby in Her Vagina"; in 2016, the runner-up was "Woman Arrested for Defecating on Boss' Desk after Winning the Lottery."[35] As suggested by cultural evolution researcher Alberto Acerbi, the most implausible fake news stories, whether or not they are political, spread largely because they are entertaining rather than because they offer justifications for anything.[36] The most absurd political fake news stories might also owe their appeal precisely to their over-the-top nature, as they make for great burning-bridges material (see chapter 12).

WHENCE POLARIZATION?

When a piece of information is seen as a justification, we can afford to evaluate it only superficially, as it will have little or no influence on what we believe or do—by virtue of being post hoc. This being the case, however, we should observe no changes at all, not even a strengthening of views. After all, a strengthening of our views is as much of a change as a weakening, and should require equally strong evidence. Yet it has been regularly observed that piling up justifications reinforces our views and increases polarization. In an experiment, people had to say how much they liked someone they had just listened to for two minutes.[37] This confederate appeared, by design, either pleasant or unpleasant. Participants who had to wait a couple of minutes before rating the confederate provided more extreme evaluations than people who answered immediately after hearing the confederate speak. During these extra minutes, participants had conjured up justifications for their immediate reaction, making it more extreme.[38]

A similar tendency toward polarization has been observed in discussion groups. In a study, American students were first asked their stance on foreign policy.[39] Doves—people who generally oppose military intervention—were put together in small groups and asked to discuss foreign policy. When their attitudes were measured after the exchange, they had become more extreme in their opposition to military intervention. Experiments that look at the content of the discussions taking place in like-minded groups show that it is chiefly the accumulation of arguments on the same side that leads people to polarize.[40]

It seems clear from the preceding that justifications for beliefs we already hold aren't always inert. Whether they are self-generated or provided by people who agree with us, they can push us toward more extreme versions of the same beliefs. Why?

When we evaluate justifications for our own views, or views we agree with, our standards are low—after all, we already agree with the conclusion. However, that doesn't mean the justifications are necessarily poor. In our search for justifications, or when we're exposed to the justifications of people who agree with us, we can also stumble on good reasons, and when we do, we should recognize them as such. Even if the search process is biased—we're mostly looking for justifications that support what we already believe—a good reason is a good reason, and it makes sense to change our minds accordingly. For instance, a mathematician convinced that a conjecture is correct might spend years looking for a proof. If she finds one, her confidence in the conjecture should be strengthened, even if the search process was biased (as she was looking to prove her conjecture, not refute it).

In the mathematician's case, there can be no polarization, only an increase in confidence: the proof supports exactly the conjecture, not a stronger version of it. By contrast, everyday arguments are much less precise: they point in a general direction rather than to an exact conclusion. Most arguments, say, against capital punishment, are not arguments for a specific position—that the death penalty should be legal only in such and such cases, with such and such exceptions. Instead, they tend to be arguments against the death penalty in general—if it is state-sanctioned murder, then it is a reason to get rid of the death penalty across the board. Piling up such arguments can lead not only to an increase in confidence but also to polarization—a stronger opposition to the death penalty in this case.

Polarization does not stem from people being ready to accept bad justifications for views they already hold but from being exposed to too many good (enough) justifications for these views, leading them to develop stronger or more confident views. Still, if people have access to a biased sample of information, the outcome can be dire.

Many commenters have linked the perceived increase in political polarization (in the United States at least) with the rise of social media. In this prevalent narrative, social media feed us a diet of news and opinions that agree with us—because we tend to follow people with the same political leanings, or because the algorithms that decide what news we see have adapted to our preferences, creating so-called echo chambers.[41] Legal scholar Cass Sunstein even devoted a book to the issue: *#Republic: Divided Democracy in the Age of Social Media*.[42]

Could it be, then, that an eagerness to justify our views, and social media providing us with an endless source of reinforcing justifications, are responsible for what is often described as one of the worst problems in contemporary U.S. politics?[43] As the title of an article in *Wired* put it, "Your Filter Bubble Is Destroying Democracy."[44] In this light, could fake news, and partisan news more generally, not be the mostly innocuous providers of post hoc justifications I have portrayed, but a grave threat to our political system?

HOW POLARIZED ARE WE? (WELL, AMERICANS, REALLY)

There might be a nugget of truth in the narrative of social media–fueled polarization, but it is hidden under layers of inaccuracies and approximations.

First, the degree of polarization is often exaggerated (throughout this section, I mostly discuss the U.S. case, which is the best studied). As political scientist Morris Fiorina and his colleagues point out, the proportion of independents (people who are neither Republicans nor Democrats) hasn't decreased in decades; if anything, it has increased in recent years, rising to 42 percent of the population in 2017.[45] Likewise, most Americans think of themselves as moderate, rather than conservative or liberal, a proportion that has remained roughly constant over the past forty years.[46] Moreover, a large majority of Americans think that Republican and Democrat politicians should compromise, and this is still true for a plurality of those with the most consistently liberal or conservative views.[47] Finally, on most issues only a minority of respondents hold extreme views: for instance, little more than 10 percent of Americans polled said that there should be no restrictions on gun ownership, or that only law enforcement officers should have guns (an extreme opinion in the United States).[48]

This lack of polarization can even be observed in the behavior of the vast majority of social media users.[49] On Twitter, the 1 percent most active users behave according to the polarization narrative, overwhelmingly sharing content that supports their political stance. By contrast, the other 99 percent tend to depolarize the informational environment: the content they share is, on average, more politically moderate than the content they receive.

The impression of increased polarization is not due to people developing more extreme views but rather to people being more likely to sort themselves consistently as Democrat or Republican on a range of issues.[50] This increased sorting is in part the outcome of people becoming better informed about where Democrats and Republicans stand on key issues.[51] In 2000, barely half of Americans understood that presidential candidate Al Gore was to the left of his opponent George W. Bush on a range of central issues, such as what the level of government spending should be. In 2016, three-quarters could say that Clinton was to the left of Trump on these same issues.[52] The only reliable increase in polarization is in affective polarization: as a result of Americans more reliably sorting themselves into Democrats and Republicans, each side has come to dislike the other more.[53]

But if social media are trapping people into echo chambers, why do we not observe more ideological polarization? Because the idea that we are locked into echo chambers is even more of a myth than the idea of increased polarization.[54] If anything, social media have boosted exposure to different points of view. After all, Facebook users are regularly exposed to more "friends" there than offline, allowing them to see the opinions of people they would barely ever talk to. An early study by economists Matthew Gentzkow and Jesse Shapiro found that the "ideological segregation of online news consumption is low in absolute terms" and "significantly lower than the segregation of face-to-face interactions."[55] Another study similarly found "no support for the idea that online audiences are more fragmented than offline audiences."[56] Still another observed that "most people across the political spectrum have centrist [online] media diets."[57]

Research conducted in other countries reaches the same conclusions. In Germany and Spain, "most social media users are embedded in ideologically diverse networks," and in the United Kingdom, "only about 8% of the online adults ... are at risk of being trapped in an echo chamber."[58] By and large, people on social media mostly look at traditional, middle-of-the road news outlets; when they are exposed to extreme views, these views tend to come from both sides of the political spectrum.

Economist Hunt Allcott and his colleagues recently conducted a large-scale experiment to test the effects of Facebook on political polarization.[59] They paid thousands of Facebook users to deactivate their account for a month and compared these users to a control group that kept on using Facebook. People who kept using Facebook did not develop more polarized attitudes and did not become more likely to support candidates from their favorite party. However, on a number of ideological measures, they were more likely to sort themselves consistently as Republicans or Democrats. On the other hand, the people who stopped using Facebook were less well informed about the news, and less likely to have seen news "that made them better understand the point of view of the other political party." Another study found that increased Facebook use was related to *de*polarization: as people were exposed to views different from their own, they developed weaker attitudes.[60] These outcomes are in line with the observation that whatever polarization might be taking place in the United States is more apparent among older than younger adults, with the former being less likely to use social media.[61]

Then, the puzzle should surely be: Why don't we observe more echo chambers and polarization? After all, it is undeniable that the internet provides us with easy ways to find as many justifications for our views as we would like, regardless of how crazy these views might be (see how many arguments in favor of flat-earth theory you can find online). However, the desire to justify our views is only one of our many motivations; usually, it is far from being a paramount goal. Instead, we're interested in gathering information about the world, information that most of the people we talk to would find interesting and credible. Even when looking for justifications, most of us would have learned from experience that simplistic rationalizations won't fly with people who do not share our point of view.[62]

WHAT TO DO?

The main message of this chapter appears to be good news. Many misguided or wicked beliefs—from the humoral theory of disease to fake news—are much less consequential than we think. As a rule, these beliefs do not guide our behaviors, being instead justifications for actions we wanted to perform anyway. On the one hand, this is good news indeed, as it means that people are not so easily talked into doing stupid or horrible things. On the other hand, this is bad news, as it means that people are not so easily talked *out of* doing stupid or horrible things. If a belief plays little causal role in the first place, correcting the belief is also unlikely to have much of an effect.

The fact that people all over the world practiced bloodletting without having ever heard of humors suggests that had the humoral theory been soundly refuted earlier in the West, it wouldn't have stopped people from wanting to be bled. It was only when evidence actually showed bloodletting didn't work that doctors stopped advocating for it. If a skeptic were to challenge gossip about the local Jewish population, more rumors would likely pop up as long as people were keen on going on a rampage, using ugly tales to scapegoat their Jewish neighbors. A refutation from the authorities might work not because it would be more convincing but because it would signal a lack of willingness to tolerate the violence. Crowds are calculating enough: in Kishinev, they paid attention to subtle signals from the police that they wouldn't interfere with the pogrom.[63]

Likewise, refuting fake news or other political falsehoods might be less useful than we would hope. As a study mentioned earlier in the chapter suggests, even people who recognized that some of their views were mistaken (in this case, some of Donald Trump's untrue statements they had accepted) did not change their underlying preferences (voting for Trump). As long as the demand for justifications is present, some will rise to fulfill it. Before the internet made fake news visible for everyone to gloat at its absurdity, it could be found in the pages of specialized newspapers—such as the canards of eighteenth-century France—with exactly the same patterns as those observed now. Most of the time, the news was pure sensationalism: one of these canards announced the discovery in Chile of a creature with "the head of a Fury, wings like a bat, a gigantic body covered in scales, and a dragon-like tail."[64] But when people wanted to give voice to their prejudices, the canards obliged, for instance, by inserting Marie Antoinette's head in lieu of that of the Fury to please the revolutionary crowds. And if newspapers couldn't do it, word of mouth would. Each individual piece of fake news would undoubtedly reach a smaller audience that way, but more would be created—witness, for instance, the rumors about nobles restricting the grain supply that emerged independently in countless villages as the French Revolution unfolded.[65]

Even if debunking beliefs that spread as post hoc justifications appears a Sisyphean task, the efforts are not completely wasted. People do care about having justifications for their views, even if they aren't very exigent about the quality of these justifications. As a decision or opinion is made increasingly hard to justify, some people will change their minds: if not the most hard-core believers, at least those who didn't have such a strong opinion to start with—which is better than nothing.

14

SHALLOW GURUS

THE CHRISTIAN GOD IS AN OMNISCIENT, omnipotent, omnipresent being, who loves all regardless of their faults. Other Christian beliefs vary as a function of the specific church one belongs to. Trinitarians believe that god is one, but also three in one: the Father, the Son, and the Holy Spirit. Catholics believe that the bread and wine of the Eucharist become the body and blood of Christ (transubstantiation). Lutherans claim, by contrast, that the bread and wine of the Eucharist adopt a dual essence, keeping their material identity while also becoming the body and blood of Christ (consubstantiation). Other forms of Christianity, from Calvinism to Methodism, adopt yet further variants of these views.

Scientists routinely defend notions that would otherwise appear equally strange. It might seem as if you're currently immobile (or moving at a low speed on a train, say), but in fact you're moving at more than 600 miles per hour (the earth's rotation), plus 67,000 miles per hour (the earth's revolution around the Sun), plus 514,000 miles per hour (the solar system's revolution around the galactic center), plus 1.3 million miles per hour (the Milky Way's movement across space). You, along with the rest of life on Earth, are descended from unicellular organisms. Tectonic plates—the big rocks weighing up to 10^{21} kilograms on which we stand—are constantly shifting. When you take a plane, time slows down because of your speed, but accelerates because of your altitude. The list is endless, through quantum superposition to the Big Bang, but the point is easily made: before they were established, many, maybe most, scientific theories would have sounded nuts to everybody but their creators.[1]

Some of the most influential intellectuals of the twentieth century were notoriously obscure writers. Until 1998, the Bad Writing Contest would select every year a scholar on the basis of the abstruseness of their prose.[2] The last person to win the first prize was philosopher Judith Butler, yet she is just one of the many Derridas, Kristevas, Baudrillards, and other (formerly) fashionable intellectuals known for their opaque prose. My personal favorite is Jacques Lacan, a French psychoanalyst who makes the most abstruse postmodern scholar look like a model of clarity. Here is an excerpt picked pretty much at random from his latest published seminar:

> To cut a long story short, I would say that nature's specificity is to not be one, hence the logical process to broach it. By the process of calling nature what you exclude from the mere fact that you are interested in something, this something that distinguishes itself from being named, nature doesn't risk anything but to affirm being a potpourri of non-nature.[3]

DEGREES OF COUNTERINTUITIVENESS

What do these very diverse ideas, from the Trinity to plate tectonics to Lacan's musings, have in common? First, they have proved at least somewhat influential, at most widely culturally successful. Across the world, around 2.4 billion people share the Christian faith. Belief in the god of the Bible is accepted by

56 percent of Americans (as of 2018).[4] Most people in rich countries trust science to a significant extent and accept the majority of the theories scientists agree on (with the odd but worrying exception).[5] Obviously, Lacan could never claim the same reach, but his authority ran deep, and he boasted many distinguished intellectuals as his groupies. Twenty years after his death, Lacan's teachings were still influential, in France at least—I should know, having had to suffer through them when I started my BA in psychology. More generally, postmodern thinkers held center stage in the Western intellectual world for a good chunk of the twentieth century and exert their influence to this day. Bruno Latour, who used to be one of them, now bemoans that "entire Ph.D. programs are still running to make sure that good American kids are learning the hard way that facts are made up, that there is no such thing as natural, unmediated, unbiased access to truth; that we are always prisoners of language."[6]

Besides their popularity, these ideas share another trait: they do not fit with our intuitions. They either challenge them or pass them by altogether.

Concepts can be more or less intuitive.[7] Take the concept of "human." Once we categorize an agent as human, we can make a wide variety of inferences: that this agent perceives things, forms beliefs, has desires and overcomes obstacles to fulfill them, likes some people more than others, needs to eat and drink, has a material body, has ancestors who were also humans, eventually dies, and so forth. Because these inferences come naturally, the concept of "human" is intuitive.

Some ideas fail to tap into any of our intuitive concepts: they are essentially incomprehensible. "Something that distinguishes itself from being named" doesn't trigger any concept I've mastered, and it fails to ring any inferential bells.

Other concepts yet go against the grain of our intuitions.[8] For instance, because we don't have well-worked-out concepts of supernatural entities, we have to rely on our concept of human, even though supernatural beings by definition violate a number of our intuitions. Ghosts are a kind of human that can walk through walls. Zeus is a kind of human who is immortal and shoots lightning bolts. The Christian god is a kind of human who is also an omnipresent, omnipotent, omniscient, all-loving being. All of these concepts are, in some ways, counterintuitive.

Religious concepts are often counterintuitive, but not all to the same degree. Pascal Boyer has argued that the vast majority of concepts of supernatural agents found across the world are only minimally counterintuitive.[9] For example, Zeus violates some of our assumptions about human agents—he is immortal, for one. But he still respects most of our preconceptions: he perceives things through his senses, forms beliefs, has desires and overcomes obstacles to fulfill them, likes some people (or gods) more than others. Likewise, ghosts are immaterial, but they still perceive things through their senses, and so forth.

By contrast, the Christian god, in his full theological garb, violates just about every assumption we have about humanlike agents. Not only is he immortal and immaterial, but he doesn't perceive things through his senses or form beliefs (he already knows everything), he doesn't need to overcome obstacles (he can do everything he wants), and he doesn't prefer some people to others (he loves everyone).

Much like the theologically correct Christian god, many scientific concepts are full-on counterintuitive. Our concept of what moving entails—the feeling that we're moving, movements of air, and so forth—is violated by the idea that we're barreling through space at a tremendous speed. Our naive sense of biology tells us that like begets like, and that microorganisms definitively don't beget humans. Our naive sense of physics suggests that humongous rocks don't drift about with no apparent cause.

To be accepted, ideas that don't tap into our intuitive concepts, or that go against them, face severe obstacles from open vigilance mechanisms. We have no reasons to accept ideas we don't understand, and we have reasons to reject counterintuitive ideas. When we engage in plausibility checking, we don't tend to reject only ideas that directly clash with our previous views but also ideas that don't fit with our intuitions more generally. For instance, you've probably never thought about whether there are penguins on Jupiter. Yet if I told you that some had been recently discovered, you would be skeptical: you have an intuition that no animals, and especially no terrestrial animals, would be found there.

Open vigilance also contains mechanisms to overcome plausibility checking and accept beliefs that clash with our previous views or intuitions: argumentation and trust.

Argumentation is unlikely to play a significant role in the wide distribution of incomprehensible ideas or counterintuitive concepts. Argumentation works because we find some arguments intuitively compelling. This means that premises and conclusions must be linked by some intuitive inferential process, as when someone says, "Joe has been very rude to many of us, so he's a jerk." Everyone can understand how being repeatedly rude entails being a jerk. But if a proposition is incomprehensible, then it can't properly be argued for. That's probably why Lacan asserts, rather than argues, that "nature's specificity is to not be one."[10]

Argumentation plays a crucial role in the spread of counterintuitive religious and scientific concepts, but only in the small community of theologians and scientists who can make enough sense of the arguments to use and construct them. Beyond that, few people are competent and motivated enough to properly evaluate the technical defense of the Christian god's omnipotence, or of relativity theory. For example, most U.S. university students who accept evolution by natural selection don't understand its principles properly.[11]

PRECIOUS SHALLOWNESS

If argumentation can't explain the widespread acceptance of incomprehensible or counterintuitive beliefs, then it must be trust. Trust takes two main forms: trust that someone knows better (chapter 5), and trust that they have our best interests at heart (chapter 6). To really change our minds about something, the former kind of trust is critical: we must believe that someone knows better than we do and defer to their superior knowledge.

The preceding examples suggest that people are often so deferential toward individuals (Lacan), books (the Bible), or specialized groups (priests, scientists) that they accept incomprehensible or counterintuitive ideas. From the point of view of open vigilance, the latter is particularly problematic. Accepting counterintuitive concepts, concepts that could wreak havoc with our cognitive systems, seems eminently dangerous, as it would involve letting other people play around with our way of thinking. For example, believing that an agent can have the properties of the Christian god could jeopardize our ability to draw inferences about humans more generally—after all, our assumptions about humans are quite sound, and it would be a shame if something happened to them.

Experiments have shown that, in fact, counterintuitive concepts do not have much of an influence on our intuitive way of thinking. In the religious domain, psychologist Justin Barrett has shown that many Christians abide by a form of "theological correctness," but that their theologically correct beliefs have little impact on how they actually think about god.[12] The Christians Barrett interviewed were able to describe god's canonical features—he knows everything, is everywhere, and so forth.[13] However, when praying, they saw god "like an old man, you know, white hair," even though they "know that's not true."[14] Moreover, when asked to retell a story about god intervening to save a drowning child, many described god's actions as sequential: first, he finishes answering one prayer, then he turns his attention and powers to the child.[15] Omniscient and omnipotent beings aren't supposed to get busy or distracted.[16]

This doesn't mean Christians can't draw inferences from their theologically correct views. If asked whether god is omnipresent, and then whether god is in both this room and the next, they would answer "yes." Still, Barrett's observations suggest that the acceptance of counterintuitive ideas remains shallow: we can assent to them, even draw inferences from them when pushed, but they do not affect the way we think intuitively. On the contrary, our intuitive way of thinking tends to seep into how we treat counterintuitive concepts, as when Barrett's participants implicitly thought that god had a limited attention span.

The same logic applies to scientific concepts. Psychologist Michael McCloskey and his colleagues were among the first to systematically investigate students' *intuitive physics*: how they answer simple physics problems intuitively, without having recourse to the explicit knowledge of physics acquired in the classroom.[17] One of the experiments involved students at an elite U.S. university, most of whom had taken some physics classes. McCloskey and his colleagues confronted the students with a series of problems, such as the one illustrated in figure 4.

Fewer than half of the students were able to provide the correct answer, namely, that the ball goes on in a straight line. Most

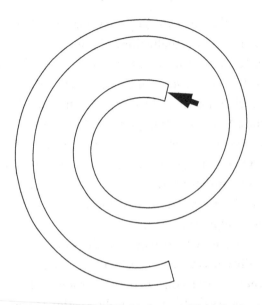

FIGURE 4. What path does a ball launched at the arrow follow when it exits the tube?
Source: Redrawn from McCloskey, Caramazza, & Green, 1980, p. 1139.

said it would keep going in a curve. This means that fewer than half of the students were able to apply the understanding of inertia they had acquired in school, according to which, in the absence of any force exerted on them, objects move in a straight line at constant speed. This notion of inertia is counterintuitive: for example, our experience tells us that objects stop moving seemingly of their own accord, absent the application of any force (a ball eventually stops rolling even if it doesn't hit a wall). The counterintuitiveness of the correct notion of inertia means that it is easily overridden by the students' intuitions about object movement. If that seems unfortunate, we should on the whole be thankful for the limited cognitive influence of counterintuitive scientific concepts. If our brains could truly process the idea that we're darting across space at tremendous speeds following complex curves, we would constantly suffer from motion sickness.[18]

These observations show that counterintuitive ideas, even when they are held confidently, have no or very limited impact on the functioning of the intuitive systems with which they are at loggerheads. To some extent, counterintuitive ideas are processed like incomprehensible ideas: even though, in theory, they should constantly clash with our intuitions, in practice they simply pass them by. Like many of the misconceptions we have explored in the last chapters, they remain reflective, detached from the rest of our cognition.

CHARISMATIC AUTHORITY?

The cognitive shallowness of counterintuitive ideas mitigates the challenge they raise for open vigilance, as accepting such ideas is much less risky than it would be if they had affected our intuitive cognitive mechanisms. But this shallowness doesn't explain why people would accept a bunch of bizarre beliefs, some of which clash with their intuitions: it still seems that people are often unduly deferential, seeing some authorities as more knowledgeable than they really are (except for scientists, whose knowledge, if anything, is likely underestimated).

A common explanation for this undue deference is that some people are charismatic: their attitude, their voice, their nonverbal language make them uniquely enthralling and even credible. Anthropologist Claude Lévi-Strauss wasn't a Lacan groupie, and yet he described "the power, the hold over the audience that emanated both from Lacan's physical person and from his diction, his movements."[19] Lacan's sycophantic French Wikipedia page even claims that his "style of discourse" "irrevocably affected" the French language.[20]

When it comes to widespread religious or scientific beliefs, charisma cannot be the main explanation. None of our Christian contemporaries have met Jesus, and I've managed to accept the concept of inertia without meeting Galileo. I don't think that personal charisma explains at all why some people are deemed more credible than others. Instead, I outline three mechanisms that lead some individuals to be perceived as more knowledgeable than they are, making their audience unduly deferential. I believe that the spread of incomprehensible and counterintuitive beliefs largely stems from a mix of these three mechanisms.

REPUTATION CREDIT

To understand why we sometimes end up thinking some people more knowledgeable than they really are, we must go back to the cues we use to deem individuals more or less knowledgeable. One of the main cues we rely on is past performance. Someone who is able to consistently fix computers is deemed competent in this area, and we are more likely to believe them when they advise us on how to fix our stalled PC. Past performance doesn't comprise only actions but also words. People who give us valuable information are deemed more competent, which leads to the question of how we decide what information is valuable.

In many cases, we can judge whether a piece of information is valuable after the fact: Did our friend's advice help us fix our computer? In other cases, we deem a piece of information potentially valuable, and think its source competent, before being sure that the information is really valuable. We give a kind of reputation credit. For information to be deemed valuable, it must be both plausible and useful.[21] For example, information about threats has the potential to be very useful, as it can help us avoid significant costs. In a series of experiments, Pascal Boyer and psychologist Nora Parren showed that people who transmit information about threats, by contrast with other types of information, are seen as more competent.[22]

Attributing competence on credit, before we're sure whether a message is actually useful or not, works well on the whole, but there are some loopholes. For instance, we might tend to overestimate the usefulness of threat information, deeming it relevant even when we have few chances of ever being exposed to the actual threat. In Boyer and Parren's experiments, one of the stories given to participants mentioned the risk of encountering leaches when trekking in the Amazon, a situation few participants would ever face. This meant not only that the information was not all that useful but also that the participants would never find out whether or not it was accurate. Indeed, this is a general problem with threat information, since, if we take the threat seriously, we should avoid it, and thus never figure out if it was a genuine threat: I've never found out whether or not any "Danger High Voltage" sign was accurate. The attribution of competence to people who circulate threats is—as was suggested in chapter 10—one of the main reasons people spread false rumors, many of which mention some threat.

Besides threats, there are other types of information that can be deemed useful without ever being seriously tested, such as justifications. Someone who provides justifications for actions people want to engage in anyway can be rewarded by being seen as more competent. However, this reputation credit can be extended indefinitely if the actions are never seriously questioned, and the justifications never tested.

This loophole in the way we attribute competence is, in the vast majority of cases, of little import. Maybe we will think a friend who warns us about the dangers of such and such exotic food a bit more competent than they actually are, but we have other ways of more accurately estimating our friend's competence. The real problem dawns with the rise of specialists: people whom we don't know personally but through their communications in a specific domain.

Nowadays, some news sources specialize in the provision of threatening information. A prime example is the media network of conspiracy theorist Alex Jones: the InfoWars website, radio show, YouTube channel, and so forth. Most stories on the InfoWars front page are about threats. Some are pretty generic threats: a lethal pig virus in China that could strike humans, a plane pilot on a murder-suicide mission.[23] Many stories are targeted: migrants from Islamic countries are responsible for most sex crimes in Finland, Turkey "announces launch of worldwide Jihad," Europe is committing suicide by accepting Muslim migrants.[24] Even a non-directly threat-related piece on George Soros's fight with the Hungarian government is accompanied by a video warning against the dangers of powerful communists such as Barack Obama (!), Richard Branson (?!), and Jeff Bezos (??!!).[25]

Presumably, few in Jones's audience live in Finland, or in close proximity to sick Chinese pigs. As a result, the readers are unlikely to find out whether the threats are real, and Jones can keep the reputation credit he earned from all these warnings. He is then able to leverage this perception of competence in a variety of ways, for instance, by selling expensive yet useless nutritional supplements whose very names remind the reader of constant threats—"Survival Shield X-2 - Nascent Iodine"—or a variety of "preparedness" products, from emergency survival food (one-year supply!) to radiation filters.[26]

Turning to justifications, we observe a similar dynamic in the case of Galen. The Roman physician provided a complex theoretical apparatus as a rationale for the relatively intuitive practice of bloodletting. Doing so made him appear more competent (note that there were other, better reasons to deem Galen competent). As a result, physicians might have heeded Galen's advice even when it departed from the most intuitive forms of bloodletting. For instance, he advocated bloodletting for a far wider range of ailments than was usual.[27] He also had rather idiosyncratic views on which veins should be cut open—the thumb of the left hand to cure the spleen, say—when in most cultures bloodletting is practiced near the ailing body part (e.g., on the temple for headaches).[28] On one point Galen appears to have been particularly influential: the quantity of blood drained. My reading of the anthropological and historical literature suggests that, in most times and places, only a tiny amount of blood was removed through bloodletting. By contrast, Galen boasts of sometimes draining two liters of blood from his patients, bleeding them until they faint.[29] More or less directly, Galen's recommendation may well have precipitated an untold number of deaths, including that of George Washington, who was bled 2.5 liters (84 ounces) before he died.[30]

On a much larger scale, there may be a similar dynamic affecting religious creeds, with the search for justification bringing in its wake an assortment of weird beliefs. Cognitive scientist Nicolas Baumard and his colleagues have argued that many of the teachings of the great world religions are intuitively appealing, at least for people in the right environment.[31] In their model, as the material environment becomes less of a constant and immediate threat, people start yearning for different moral norms that emphasize "moderation, self-discipline, and withdrawal from excessive greed and ambition."[32] Leaders emerge who promote religious justifications for these new norms, with god(s) that care about human morality, and a world imbued with cosmic justice. This is a remarkable departure from previous religious worldviews where, for instance, Zeus and his ilk displayed no superior sense of morality. The leaders who are able to articulate religious creeds fitting these changing moral intuitions are rewarded with deference. One of the effects of this deference, arguably, is to help spread other ideas, ideas born of the religious specialists' striving for a more intellectually coherent system. These ideas don't have to be particularly intuitive, or to be of any use as justifications for most of the flock. For example, few Christians care deeply about what happens to the soul of people who lived before Jesus and thus couldn't be saved by the sacraments (the unlearned). Yet theologians had to ponder the issue, and made this part of the official creed: for example, in Catholicism the unlearned are stuck in the "limbo of the fathers" until the Second Coming.[33] More significantly, the theologically correct version of the Christian god—the omni-everything version—is the result of a slow elaboration over the ages of scholars attempting to reconcile various doctrines.[34]

This account distinguishes two broad sets of beliefs within the creeds of world religions. The first set comprises beliefs that many people find intuitively compelling—for example, rewards and punishments in the afterlife for good and bad deeds. The second comprises beliefs that are relevant only to the theologians' attempts at doctrinal coherence. We find both categories in world religions besides Christianity. Crucially, the first set of beliefs is quite similar in every world religion, while the second varies widely. For example, in Buddhism the concept of merit plays a central role, so that those who do good deeds have better luck in their next lives. But we also find in Buddhism counterintuitive ideas that play little useful justificatory role and have no parallel in Christianity, such as the precise status of the Buddha in relation to humans and gods, or the cycle of reincarnation.

TRICKLE-DOWN SCIENCE

Being willing to give people the benefit of the doubt and grant them a good reputation on credit, because they warn us about threats we will never face or provide justifications that will never get tested, cannot explain the widespread acceptance of counterintuitive scientific theories. For one thing, scientific theories are nearly all counterintuitive, so scientists can't surf on a wave of easily accepted theories to make the public swallow the rest.

Frankly, I'm not quite sure why so many people accept counterintuitive scientific theories. I'm not saying they shouldn't, obviously, merely pointing out that the popularity of such counterintuitive ideas, even if they are right, is on the face of it quite puzzling. It is true that people accept scientific beliefs only reflectively, so the beliefs interact little with other cognitive mechanisms. But, still, why accept these beliefs at all? Very few people can properly evaluate scientists' claims, especially when it comes to novel discoveries. A small group of specialists understands the content of new results, can interpret them in light of the literature, and knows the team that produced them. Everybody else is bound to rely on relatively coarse cues. The further removed we are from the group of relevant experts, the coarser the cues.

There are numerous cues people use to ascertain how "scientific" something is. One is mathematizing: if math is used, then the outcome is more likely to be thought of as good science. In an experiment conducted by psychologist Kimmo Eriksson, participants, all with postgraduate degrees, had to evaluate some social science results. Half of the participants read abstracts in which a sentence with mathematical symbols had been inserted.[35] The sentence itself made no sense, yet it boosted the positive evaluations of the abstract. Another coarse cue is proximity to the hard sciences. Psychologist Deena Weisberg and her colleagues asked participants to evaluate explanations of well-established psychological phenomena.[36] Some explanations were purposefully circular, but some of these poor explanations were supplemented with useless information about the brain areas involved. The addition of irrelevant neuroscience data made participants less critical of the useless explanations. Reassuringly, genuine experts weren't fooled either by the fancy math or by the neuroanatomical babble.

An even coarser cue is the prestige of a university. A journalist reporting on two studies, one conducted at Harvard and the other at Bismarck State College, would likely stress the former affiliation more than the latter.[37] The effect of university prestige is even visible in that most dramatic demonstration of deference toward science: the Milgram obedience experiments.

The standard narrative surrounding these experiments is that Milgram showed how two-thirds of U.S. participants were willing to shock someone almost to death, if ordered to do so.[38] These results have been taken as support for Hannah Arendt's contention, based on the behavior of many Germans in World War II, that "in certain circumstances the most ordinary decent person can become a criminal," suggesting that people would obey just about any orders from any person in a position of authority.[39] However, this narrative ought to be substantially revised in two ways.

First, the two-thirds figure is inflated. It was only obtained in one variant of the experiment. Other versions, with minor changes such as a new experimenter, yielded lower rates of compliance.[40] More important, many of the participants—nearly half of them—expressed doubts about the reality of the whole setup.[41] Those who didn't express such doubts, who presumably really thought they were shocking someone, were vastly less likely to comply: only a quarter of them went all the way to the highest voltage.[42]

Second, Milgram's experiments demonstrate only that people defer to science, not to anyone barking orders at them. The participants, most of whom had quite modest backgrounds, were invited to the prestigious Yale University, welcomed by a lab coat–wearing scientist, and given an elaborate scientific rationale for the experiment. Participants followed the experimenter's request only when they believed in the study's scientific goal.[43] By contrast, straight-up orders, such as "You have no other choice, you must go on" tended to have the opposite effect, prompting participants to rebel and refuse to take further part in the experiment.[44] Removing some of the cues that made the experiment appear more scientifically respectable—for instance, carrying it out in a generic town rather than at Yale—decreased the compliance rate.[45]

The Milgram experiment illustrates the dangers of an overreliance on coarse cues to evaluate scientific value. Other examples abound. Pseudoscientists, from creationists to homeopaths, use credentials to their advantage, touting PhDs and university accreditations they gained by professing different beliefs.[46] Still, on the whole coarse cues play a positive role. After all, they do reflect reasonable trends: mathematization vastly improves science; the hard sciences have progressed much further than the social sciences; someone with a PhD and university accreditation is likely to be more knowledgeable in their field of expertise than a layperson.

The Guru Effect

Jacques Lacan relied on these coarse cues to boost his stature. He had the proper credentials. He made extensive use of mathematical symbols.[47] Still, even knowing this, I suspect few are able to plow through his seminars, and those who do, instead of being impressed by Lacan's depth, are more likely stunned by his abstruseness. How could prose so opaque become so respected?

More obscure statements require more effort to be understood; as a result, everything else being equal, obscurity makes statements less relevant.[48] Take the following example: instead of reading, "In the event of an impact where the airbag is deployed, the inflator part of the airbag may ignite in such a manner that it creates excessive internal pressure. As a result the metal inflator casing may rupture, causing metal fragments to be propelled through the airbag and into the vehicle," people would rather be told, "Your airbag might explode and kill you with shrapnel" (yes, this is a real example).[49] As a rule, when hard-to-understand content spreads, it is not because it is obscure but in spite of being obscure, when there is no easier way to get the content across.

Yet the success of Lacan, and other intellectuals of his ilk, suggests that obscurity sometimes helps, to the point that people end up devoting a lot of energy to deciphering nonsensical statements. Dan Sperber has suggested that, in unusual circumstances, obscurity can become a strength through a "guru effect."[50]

Imagine Lacan in 1932. He has attended the best schools; has been mentored by the best psychiatrists; and his noted doctoral dissertation reflects a broad mastery of the psychiatric, psychoanalytic, and philosophical literatures. He promotes the idea that mental illnesses are not necessarily deficiencies but merely different ways of thinking, which should be understood in their own terms.[51] His thesis might be right or wrong, but it is understandable, controversial, and interesting. Lacan makes a name for himself in Parisian intellectual circles, where he is, for broadly justifiable reasons, perceived as an expert on the affairs of the mind.

To maintain his status, Lacan should keep developing new and interesting theories about the mind. But this is rather difficult (believe me on this). Fortunately, there is a way out. He can rely on increasingly vague concepts, concepts that were already part of the zeitgeist. Here is an excerpt from a presentation Lacan gave in 1938: "The first case [a patient] shows how the symptoms were resolved as soon as the oedipal episodes were elucidated, thanks to a nearly purely anamnestic evocation."[52] It takes a bit of effort, and some familiarity with psychoanalytic jargon, but it is possible to make sense of this statement, which broadly says, "The patient's symptom subsided when he was able to remember having sexual desire for his mother" (a likely mistaken conclusion, but that's another issue).

Lacan's work confirms his mastery of the most complex psychoanalytic theory and suggests that decoding his dense prose is worth people's while. Because they assume Lacan to be an expert, his followers devote growing amounts of energy and imagination to make sense of the master's pronouncements. At this stage, the vagueness of the concepts becomes a strength, giving Lacan's groupies leeway to interpret his ideas in myriad ways, to read into the concepts much more than was ever intended. As noted by two of his detractors, "Lacan's writings became, over time, increasingly cryptic . . . by combining plays on words with fractured syntax; and they served as a basis for the reverent exegesis undertaken by his disciples."[53]

Still, had Lacan been followed by isolated individuals, in the absence of any external indication that their growing efforts would be rewarded, most would likely have given up long before things had reached the heights of Lacan's later years. But the groupies were, as the name suggests, a group, seeing in the others' efforts an affirmation of their own interpretive labors. As Lévi-Strauss noted when he attended one of Lacan's seminars: "I found myself in the midst of an audience that seemed to understand."[54]

Once it is widely assumed the master's edicts unfailingly reveal deep hidden truths, any admission to the contrary is seen as either an intellectual failure—that one is too dense to fathom the "crystal-clear" prose—or, worse, as an act of treason warranting ostracism. The guru even raises the stakes by proclaiming the transparency of his discourse, as when he states: "In simple terms, this only means that in a universe of discourse nothing contains everything."[55] If this is so simple, then those who don't understand must really be dunces. And so the followers opine: "Lacan is, as he himself says, a crystal-clear author."[56] Members of the inner circle cannot admit that the emperor is naked, thus preserving the illusion that Lacan's obscurity hides profound revelations.

To make things worse, the pupils are credentialed, forming the next generation of public intellectuals and university professors. This greatly extends the master's influence, as outsiders are bound to wonder how such a group of smart people could be so utterly misguided. Again, obscurity plays in Lacan's favor. If his theories were understandable, outsiders could form their own opinions. But their obscurity protects Lacan's writings from the prying eyes of critics, who must defer to those who seem to be knowledgeable enough to make sense of it all, or reject them en bloc and risk looking as if they have no appreciation for intellectual sophistication.

WHAT TO DO?

On the whole, people are pretty good at figuring out who knows best. But there are exceptions. In this chapter, I have described three mechanisms through which people might end up being unduly deferential, leading them to ponder incomprehensible beliefs, endorse counterintuitive ideas, and, occasionally, inflict (what they think are) severe electric shocks on a hapless victim. I will now suggest some potential remedies to alleviate the consequences of each of these mechanisms.

The first mechanism relies on the granting of reputation on credit: thinking people competent when they say things that appear useful, but that will never be properly checked (such as Alex Jones's dire warnings). In theory at least, the solution is relatively straightforward: to stop granting so much reputation on credit. Take the case of threats. We can still pay attention to people who warn us of various threats and take what they say into consideration, but we should stop rewarding them with our deference until we have more information about the reality of the threat. The same goes for justifications. Maybe there's a pundit we enjoy in part because they always provide us with articulate rationales for our preexisting opinions. If these justifications are then properly evaluated—we use them in arguments with friends who disagree with us, say—everything is fine. But if the justifications are not tested, then it is likely we have not only accepted dubious information but also formed an inflated opinion of a particular pundit.

A second way of becoming unduly deferential is to rely on coarse cues to estimate how scientific a piece of information is, with the risk of thinking the information more scientific than it is. As mentioned earlier, there is no magic trick here, as only some experts are typically able to evaluate in depth a new scientific result. Everyone else must rely on more or less coarse cues. Still, we can all strive to use finer-grained cues. Philosopher Alvin Goldman suggested a series of cues people could use to evaluate scientific claims, from how consensual the claims are among experts, to whether the scientists who defend the claims have conflicts of interests.[57] We should in particular be wary of flashy new results, opting to rely instead on work grounded in many separate studies. In the field of medicine, the Cochrane organization provides systematic reviews whose conclusions are vastly more reliable than the latest headline about coffee/wine/blueberries/kombucha causing/protecting us from cancer. In any case, we shouldn't turn our noses up at coarse cues: they might help some shady stuff spread, but they are still better than a blanket resistance to science, which seems to be the only practical alternative.

Finally, how to get rid of gurus who rely on the obscurity of their pronouncements to hide the vacuity of their thought? After all, even if Lacan, along with the great wave of impenetrable postmodern thinkers from the mid-twentieth century, is dead, gurus still walk among us. Jordan Peterson is a psychologist who has become incredibly popular, in part thanks to his intuitive defense of conservative ideas. Other parts of his oeuvre, however, are somewhat more baroque, such as this snippet from his *Maps of Meaning*:

> The constant transcendence of the future serves to destroy the absolute sufficiency of all previous historically determined systems, and ensures that the path defined by the revolutionary hero remains the one constant route to redemption.[58]

While we certainly haven't reached terminal Lacanianism, I still find it difficult to figure out what any of this means (even in context). The equally popular Deepak Chopra is also known for his enigmatic tweets, such as "Mechanics of Manifestation: Intention, detachment, centered in being allowing juxtaposition of possibilities to unfold," or "As beings of light we are local and non-local, time bound and timeless actuality and possibility."[59] Fortunately, spotting gurus is comparatively easy: they have no standing in the scientific community—at least not for the part of their work for which they use their guru status. Outside of the sciences that rely heavily on mathematics (and some might argue even then), just about any idea should be communicable with enough clarity that an educated and attentive reader can grasp it. If something looks like a jumble of complicated words pasted together, even in context, and after a bit of effort, then it probably is.

Doing this work is all the more important as one of the most plausible reasons why so many people like to have a guru is that the guru allows them to look more competent and knowledgeable, as the members of Lacan's inner circle did in France. While this may not always be a conscious process, the fact that the followers of a guru tend to be so vocal about the guru's intellectual prowess and depth of wisdom suggests the process isn't one of purely individual enlightenment. By challenging this prowess and wisdom, we deprive the followers of one advantage of having a guru, and the guru of some followers.

15

ANGRY PUNDITS AND SKILLFUL CON MEN

IN HIS 2004 BOOK *The Company of Strangers*, economist Paul Seabright points out how weird humans are in their reliance on strangers, people to whom they aren't related, and, increasingly, people whom they have never met in person. Until relatively recently in our history, most of the people we cooperated with were well known to us, and we could use a long trail of interactions to gauge people's value as cooperation partners.[1] Nowadays, the situation has changed: we get our news from journalists, our knowledge of how the world works from scientists, and moral guidance from religious or philosophical leaders, often without ever meeting any of these people in person. We also let surgeons we have met only once operate on us, teachers we barely know educate our children, and pilots we have never seen fly us across oceans. How do we decide who to trust in these novel situations?

In this chapter, I explore two of the ways in which we end up trusting the wrong people. The first is when people display their loyalty to us, or to our group, by taking our side in disputes even though it does not cost them anything to do so. The second is when we use coarse cues—from someone's profession to their ethnicity—to figure out who to trust. Both mechanisms can make us trust too much—I shouldn't have believed that fake doctor who scammed me of twenty euros. Still, on the whole we are more likely to err by not trusting when we should, rather than by trusting when we shouldn't.

TAKING SIDES

Even if the problem we now face routinely—how to trust complete strangers—is evolutionarily novel, we still rely on the cognitive mechanisms that evolved to help us find allies in a very different environment. A crucial element is that our allies should have our back: When a conflict arises between us and someone else, whose side are they on? We see these moments as defining in relationships. An employee only knows if the manager is truly supportive when there's a dispute with a client. We learn the extent of a romantic partner's commitment by looking at how they behave in a conflict between us and their friends. Our colleague's allegiances are made clear when a fight erupts between cliques at work.

These moments are revealing because taking sides is costly: those we do not side with see our behavior as a clear sign that they have been spurned, and in turn see us as less desirable cooperation partners. The logic is broadly similar to that of burning bridges, except that it antagonizes only one specific individual or group instead of provoking as many people as possible. In both cases, the signal—that we want to affiliate with a given individual or group—is made credible by the costs in terms of lost opportunities to affiliate with others.

In small communities, where everybody knows everybody, this signal is indeed credible: the people we side against are people we could have cooperated with, so the costs are genuine. Indeed, the higher the costs, the more credible the signal. In the schoolyard, if you get in a fight with an unpopular wimpy kid, it doesn't cost others much to take your side. But those who support you in a fight against the school bully are risking something, and their commitment is all the more meaningful.

In our modern environments, it is quite easy to take sides without paying any costs. Imagine I'm having drinks in a bar with a friend, and he gets into a verbal argument with the people at the next table. Taking my friend's side is essentially costless, as it is unlikely I will ever see the people he's arguing with again. As a result, it is not a strong indicator of how much the friendship means to me. The strategy of appearing to take people's sides, while paying only minimal costs, is widely used by social media personalities, pundits, and even entire news channels.

A good example from the United States is that of cable news networks. For many years, news networks in the United States were broadly nonpartisan, barely taking sides, presumably to avoid antagonizing part of their audience. When Fox News Channel and MSNBC were created, they had a slight slant (to the right and the left, respectively), but they mostly stuck with the same plan. However, their strategy shifted over the years, as they increasingly relied on market fragmentation to gain audience share.[2] Instead of attempting to please everyone, Fox News Channel targeted conservative Republicans, MSNBC liberal Democrats. Both channels increased their slant, so that it became quite transparent who they were siding with. However, while these channels—and their hosts—pay a cost for their partisanship, it pales by comparison with the benefits: what they lose in terms of viewers from the other side is more than made up for by gains from the side they cheer for. In this sense, both cable news networks (and the many other players who rely on the same strategy) hijack our cognitive mechanisms. They take our side in what we perceive to be cultural battles with people on the opposite side of the political spectrum. But they do so while paying only a small cost in lost audiences, so their stance does not reveal any genuine commitment.

To make things worse, the strategy of taking sides to win over an audience encourages the spread of misrepresentations about the power of our (supposed) enemies, or the very existence of these enemies. As noted earlier, the degree of commitment signaled by the act of taking sides depends on the costs incurred, and thus, inter alia, on the power of those we side against. Agents who want to gain our trust by taking our side thus benefit from portraying the other side as immensely powerful. Fox News says liberals control the media, political discourse, the universities. MSNBC claims conservatives control most political offices, big businesses, financial contributions. Some of these portrayals are more accurate than others, but they all underestimate the various countervailing forces, checks and balances, that foil the ambitions of even the most powerful groups on either side. Still, these portrayals are sure to find an avid audience, as information about the power of other groups is deemed highly relevant. At the same time, the complexity of our economic and political environments is all too easily ignored by cognitive mechanisms that evolved by dealing with much simpler coalitions.

An even more fundamental prerequisite for the strategy of taking sides is that there should be sides to begin with. While we're all embroiled in a variety of low-grade disputes between groups—with family members, neighbors, colleagues—these are too local to be of any interest to, say, a cable news channel. Instead, the conflicts must involve as many individuals as possible: on our side, so that the channel gains more audience, and on the other, so that the enemy looks more powerful. Agents, such as hosts on cable news networks, who rely on the taking-side strategy to gain audiences, benefit if they portray the world as divided and polarized.

As we saw in chapter 13, U.S. citizens are not all that ideologically polarized. However, they are perceived as being so: several studies observed that "people significantly misperceive the public to be more divided along partisan lines than it is in reality."[3] For example, the attitudes of Democrats and Republicans on free trade are remarkably similar, being very close to the middle of the road, with a slightly more positive view for Republicans. However, Democrats are perceived as being anti–free trade (which they aren't, on average), and Republicans as being strongly pro–free trade (which they aren't, again on average). These mistaken perceptions are driven by news consumption.[4] In some countries, this means TV, but the most reliable driver of inflated *perceived* polarization is the heavy consumption of online media. This makes sense: a TV channel can attempt to portray the other side as made up of crazy extremists, but on social media, these crazy extremists are there for all to see, and it is easy to forget that they represent only a sliver of the population. Social media don't make us more polarized, but they make us think we are; more precisely, social media don't push their users to develop stronger views but, through increased perceived polarization, they might contribute to increased affective polarization, as each side comes to dislike the other more.[5]

When agents with wide audiences take sides, they are incentivized to create a distorted view of the coalitional stakes—making the other side appear stronger, creating conflicts out of nothing. If this strategy is successful, it can yield further epistemic distortions.

Agents that are perceived as taking our side in conflicts against a powerful enemy gain our trust: we believe they have our best interests at heart. Moreover, as they provide us with information that supports our views, they also come to be seen as competent (as explained in the last chapter). In some cases at least, this strategy works: for instance, conservative Republicans find Fox News to be more credible than CNN, historically a broadly neutral network (although things are changing with the Trump presidency).[6]

This increased trust allows the transmission of some false information, at least at the margin. Cable news networks with a political slant spread more falsehoods than more neutral networks.[7] This obviously doesn't mean all these falsehoods are believed; still, the attempt betrays an assumption by the networks that they won't be questioned. More important, the asymmetry in trust—when we trust people deemed to be on our side much more than those deemed to be on the other side—hinders the transmission of accurate information. We aren't challenged by the people we trust, and we don't trust the people who challenge us, potentially distorting what we know.

A series of clever studies have investigated the effect of Fox News Channel availability on political opinions and political knowledge. These studies rely on the fact that Fox News Channel was introduced in different U.S. towns in a somewhat haphazard fashion, as a function of deals signed with local cable companies. As a result, the researchers were able to look at the effects of Fox News availability on a range of outcomes and treat the results as if a giant randomized experiment had been conducted. These data show that Fox News Channel did have an effect on political views, making towns where it was available slightly more Republican leaning.[8] What about political knowledge? Fox News made people more selectively knowledgeable.[9] Where Fox News was available, people tended to know more about issues well covered by Fox (rather unsurprisingly), but also to know less about issues poorly covered by Fox. Fox mostly covered issues for which the Republican Party was in broad agreement with its base. As a result, viewing Fox News reinforced the impression that the Republican Party platform aligned with the viewers' opinions, strengthening support for the party.[10] Even if, in this case, the information being presented might not have been entirely fair and balanced, this example still supports Andrew Gelman and Gary King's contention that the media can affect political outcomes, but chiefly "by conveying candidates' [or the parties'] positions on important issues."[11]

While there is a danger that the hijacking of our coalitional thinking is turning the media landscape into increasingly vociferous fights between partisan hacks, it's good to keep in mind that there are countervailing forces. We can recognize that media personalities who appear to be on our side are, more often than not, of little use to us. At best, they provide us with information that justifies our views, but this information has to be sound to be truly relevant, something we only discover when we use the information in an adversarial debate. There is a social cost to be paid when we attempt to justify our views with arguments that are too easily shot down. Apart from those that cater only to extreme partisans, most media thus have an incentive to stick to largely accurate information—even if it can be biased in a number of ways.[12] Moreover, our reaction to challenges isn't uniformly negative. In a fight with our partner, we might get angry at a friend who supports our partner instead of us. But, if they make a good point that we're in the wrong, we'll come to respect them all the more for helping us see the light (although that might take a little time). We're wired to think in coalitional terms, but we're also wired to form and value accurate beliefs, and to avoid looking like fools.

TRUST IN STRANGERS

When it comes to social media personalities or news channels, at least we have time to gauge their value as information providers, as we see them on TV night after night. What about people we have only just met? How do we know whether they have our interests at heart? Given the lack of information about these strangers' past behavior, we must rely on coarse cues about their personality, the groups they belong to, and their current situation. These cues range from the very general (does this individual appear trustworthy?) to the very specific (is this individual well disposed toward me at this moment?).

As an example of a general trait, consider religiosity. In some cultures religious people are seen as particularly trustworthy.[13] As a result, in these cultures people who wear badges of religious affiliation are seen as more trustworthy even by the nonreligious.[14] By contrast, other cues indicate trustworthiness only in the context of specific relationships. In a series of experiments, students were asked to say whom they would trust to be more generous toward them: another student from their own university, or a student from another university. The participants put more trust in the students at their own university, but only if they knew the students also knew the participants belonged to the same university. The participants did not think their fellow students would be more generous as a rule, only more likely to prove generous with those sharing an affiliation.[15]

People rely on a variety of cues to decide who they can trust, from displays of religiosity to university affiliation. But how do these cues remain reliable? After all, if appearing to be religious, or to belong to the local university, makes one more likely to be trusted, why wouldn't everyone exhibit these cues whenever it could be useful? These cues are kept broadly reliable because they are in fact signals, involving some commitment from their sender, and that we keep track of who is committed to what. Someone who wears religious clothes but does not behave like a religious person will be judged more harshly than someone who behaves in the same way but does not display religious badges. In an extreme use of religious badges, Brazilian gang members who want a way out can now join a church, posting a video of their conversion on social media as proof. But this isn't a cheap signal. Members of other gangs refrain from retaliating against these new converts, but they also keep close tabs on them. When a young man posted his conversion video just in time to avoid being killed, the rival gang members "monitored him for months, checking to see if he was going to church or had contact with his former [gang] leaders."[16]

More generally, we tend to spurn people who pretend to be what they aren't. If I walked around hospitals in scrubs wearing a "Dr. Mercier" tag, people would be justifiably annoyed when I revealed that my doctorate is in cognitive science. Even a construction worker who dressed and behaved like a rich businessman would face difficulties integrating with other workers, or with rich businessmen.

Still, some people can, at least in part, get away with pretending to be who they aren't. Con men are a good example.[17] In *The Sting*, the characters played by Robert Redford and Paul Newman describe their world as that of grifters, opposed to the world of citizens, a world to which they couldn't and wouldn't want to belong. Big cons took time, as the hustlers had to progressively earn the mark's trust, to "play the con for him" (as the protagonists do in *The Sting*).[18] This involved letting the mark get to know the con men, allowing the mark to earn some money, and setting up such an elaborate story that it became a stretch to believe it was all made up. The con perpetuated in *The Sting*—inspired by real life—involved renting a room, disguising it as a betting saloon, and hiring dozens of actors to play the role of other gamblers. It is a wonder that more people did not fall for such cons.

Minor cons, by contrast, require minimal contact between the con man and the mark. The first man to be called a con man was Samuel Thompson, who operated around 1850 in New York and Philadelphia.[19] He would come up to people, pretend to be an old acquaintance, and remark on how people did not trust each other anymore. Making his point, he would wager that the mark wouldn't trust Thompson with their watch. To prove him wrong, and to avoid offending someone who appeared to be a forgotten acquaintance, some people would give Thompson their watch, never to see him or their watch again.

Thompson relied on his "genteel appearance" (a coarse cue indeed) to pressure his victims: they might not have trusted him altogether, but they feared a scene if they blatantly distrusted someone of their own social standing.[20] This is how the fake doctor from the introduction got me to give him twenty euros. Once you accept the premise that someone is who they say they are, a number of actions follow logically: had that person been a real doctor, I should have been able to trust him with the money. And rejecting the premise, saying to someone's face that we think they are a fraud, is socially awkward.

The same techniques are used in social engineering: instead of hacking into a computer system, it is often easier to obtain the desired information from a human. In *The Art of Deception*, hacker and social engineer Kevin Mitnick describes how valuable information can be extracted from employees. In one example, the social engineer calls up an employee, pretends to be from a travel agency, and makes up a phony trip that the employee supposedly booked.[21] To understand how the error might have occurred, the employee is asked to provide his employee number, which later allows the social engineer to impersonate him. Again, the employee was relying on coarse cues: that the individual on the line sounded like a genuine travel agent.

The example of con men and social engineers suggests that relying on coarse cues to trust strangers is a daft move, easily abused. In fact, conning people is harder than it seems. For one thing, we mostly hear about the cons that work. In total, six people lodged official complaints against Thompson for theft—not a huge number to start with, and we don't know how many people he had tried his luck with and failed.[22] Indeed, by all accounts he was a "clumsy thief and unsophisticated scammer."[23]

Ironically, that most egregious of cons, the 419 scam, or Nigerian scam, illustrates how hard scamming really is.[24] A few years back, we were bombarded with e-mails alerting us to a wonderful opportunity: someone, often from Nigeria, had a huge amount of money and offered us a cut of the pie if we would only wire them the small sum they needed to access a much bigger sum. This small investment would be repaid a hundredfold. Seeing these ludicrous messages, it is quite natural to think people incredibly gullible: How could anyone fall for such tall tales, sometimes losing thousands of dollars?[25] In a perceptive analysis, computer scientist Cormac Herley turned this logic on its head: the very ludicrousness of the messages shows that most people are, in fact, not gullible.[26]

Herley started by wondering why most of these messages mentioned Nigeria. This scam had quickly become associated with the country, so much so that *scam* was one of the top autocompletes after typing *Nigeria*. Why, then, keep using the same country? Besides the country, there was clearly little attempt at credibility in the messages: the sender was a prince ready to part with a good chunk of a huge sum, not exactly a common occurrence. Why make the messages so blatantly suspect? Herley noted that while sending millions of messages was practically free, responding to them cost the scammers time and energy. After all, no one would be sending the money right away. Instead, marks had to be slowly reeled in. Expending such effort was only worthwhile if enough marks ended up falling for the scam hook, line, and sinker. Anyone who would do a Google search, ask for advice, or read their bank's warning notices wouldn't be worth expending any effort on. The solution scammers adopted to weed out these people was to make the messages voluntarily preposterous. In this way, the scammers ensured that any effort spent engaging with individuals would only be spent on the most promising marks, those who were the least well informed. Ironically, if these scam attempts are so ludicrous, it is not because people are gullible but because, by and large, they aren't. If they were, scammers could cast a much broader net with more plausible messages.

Effective Irrational Trust

Not only is getting conned a relatively rare occurrence, but there is a huge benefit from relying on coarse cues to trust strangers: it allows us to trust them at all. Economists and political scientists have devised a great variety of so-called economic games to test whether people behave rationally in simple, stylized interactions. One of these is the trust game, in which one player (the investor) is provided with an initial monetary endowment. They can choose how much to invest in the second player (the trustee). The amount invested is then multiplied (typically by three), and the trustee can choose to give back any amount to the investor. To maximize the overall benefits in a fair manner, the investor would give all the money to the trustee, who would then give half of it back. However, once the investor has transferred the money, nothing stops the trustee from keeping it all. Knowing this, the investor should not transfer anything. No transfer is thus, in theory, the rational outcome. Moreover, messages from the trustee to the investor should have no effect, since they are the quintessential cheap talk: investors can promise to give back half of the money, but no extrinsic force can make them keep their word.

Yet many experiments have found that investors typically transfer a good chunk of their endowment, and that trustees tend to share back some of the proceeds.[27] Moreover, promises work. When trustees are given the opportunity to send a message to the investors, they often promise to send money back. Investors are then more likely to transfer money to the trustees, and trustees to share the money back.[28] The mere fact that someone has made a promise is sufficient to increase the level of trust, thereby generating a superior (even if, in a way, less rational) outcome. In this case, the coarsest cue—that the trustee would be a broadly similar person to the investor—is sufficient to generate some trust, and to make promises credible. This doesn't quite mean that people trust blindly, as such cheap promises lose some of their power when the stakes increase.[29]

Social scientist Toshio Yamagishi highlighted another advantage from trusting even when short-term rationality dictates we shouldn't, pointing out a fundamental asymmetry between trusting and not trusting in terms of information gains.[30] If you choose to trust someone, more often than not you'll be able to tell whether your trust was warranted. If a new classmate asks to borrow your notes and promises to give them back to you the next day, you'll only know if they'll keep their word if you lend them the notes. By contrast, if you don't trust someone, you might never know whether they would in fact have been trustworthy. If a friend tries to set you up with someone you don't know, it's only if you follow the dating advice that you'll figure out whether or not it was solid.

Admittedly, there are situations in which we can gauge the value of someone's word without having to trust them first. For instance, you can see whether investment advice pans out without following it, simply by keeping track of the relevant stocks. Still, as a rule, we learn more by trusting than by not trusting. Trust is like any other skill: practice makes perfect.

As a result of this asymmetry between trusting and mistrusting, the more we trust, the more information we gain. We not only know better which specific individuals are trustworthy but also use these experiences to figure out what kind of individual, in what kind of situation, should be trusted. In a series of experiments, Yamagishi and his colleagues found that the most trustful of their participants—those more likely to think that other people could be trusted—were also the best at ascertaining who should be trusted (in games analogous to the trust game).[31] Likewise, people who are the least trusting are the least able to discriminate between phishing attempts and legitimate interfaces.[32]

My maternal grandparents are the best illustration I know of Yamagishi's ideas. On the surface, they might seem like easy prey: they aren't so young anymore (being in their early nineties at the time of this writing), they are supernice, and are always there when a friend or a neighbor (or indeed my wife and I) need something. One doesn't get much more grandmotherly than my grandmother, plying children with sweets and giving big hugs. Yet my grandparents have a very shrewd judgment, skillfully applying selective trust. I have never seen them fall for any marketing stunt, and all their friends are perfectly trustworthy. By giving people the benefit of the doubt in initial interactions with little risk, they have accumulated a wealth of knowledge about who can be trusted and have met enough people that they could afford to select the most reliable as friends.

In spite of the informational gains that can be accrued from trusting when in doubt, the general logic of open vigilance mechanisms suggests that, on the whole, we make more errors of omission (not trusting when we should) than of commission (trusting when we shouldn't). This might seem counterintuitive, but beware the sampling bias: we're much more likely to realize we shouldn't have trusted someone when we did (we follow our friend's advice and end up on a horrible date) than to realize we should have trusted someone when we didn't (we don't follow our friend's advice and fail to meet our soul mate). The main issue with using coarse cues isn't that we trust people we shouldn't (trusting a con man because he's dressed as a respectable businessman), but that we don't trust people we should (mistrusting someone because of their skin color, clothing, accent, etc., when in fact they are perfectly trustworthy).

Experiments with economic games support this prediction. Economists Chaim Fershtman and Uri Gneezy asked Jewish participants in Israel to play trust games.[33] Some of the participants were Ashkenazi Jews (mostly coming from Europe and the United States); others were Eastern Jews (mostly coming from Africa and Asia). By and large, the former group had higher status and was expected to be perceived as more trustworthy. This is indeed what Fershtman and Gneezy observed. In a trust game, male investors transferred more money to Ashkenazi trustees than to Eastern trustees. However, the relative mistrust of the Eastern Jews was unwarranted, as Ashkenazi and Eastern trustees returned similar amounts. The same pattern was observed by economist Justine Burns in South Africa.[34] In her experiment, investors transferred less money to black trustees than to other trustees, even though black trustees then returned as much money.[35] In these experiments at least, the participants would have been better off recalibrating their coarse cues and trusting more these ethnic groups.

WHAT TO DO?

How can we better calibrate our trust? The two trust calibration mechanisms I have explored here are quite distinct and call for different adjustments. When it comes to the taking-sides strategy, we should be aware that it can be abused by people who claim to be on our side but aren't actually paying any cost for their commitment. We should be wary of largely made-up controversies with largely made-up enemies. If we base our representation of the other side on how it is portrayed in the news or, worse, on social media, then this representation is likely to be wide off the mark—mistaking, say, crazy conspiracy theorists for average Republicans, or enraged social justice warriors for typical Democrats. We must remind ourselves that the members of the "other side" are probably not that different from us, and that engaging with them is worthwhile.

What about coarse cues? When we have to rely on coarse cues—for example, when we meet someone for the first time—I believe we should try to worry less about how people judge our decisions to trust or not to trust. Con men and social engineers often rely on our reluctance to question our interlocutors, our fear of appearing rude because we don't trust them. After all, if you meet someone who really is a long-lost acquaintance, and you suggest they are trying to scam you, they will be justifiably annoyed. Not wishing to be thought ill of also drives some of our misplaced mistrust, as we're afraid of looking like fools if we get played.

In both instances we should strive to resist these social pressures. The long-lost acquaintance shouldn't put us in a situation in which we have to immediately trust them with something significant (like an expensive watch). If they do, they are the ones who are breaking social norms, not us when we refuse to grant trust under pressure. As for the fear of looking like we're easily tricked, we should strive to remember the information we gain by trusting people, even when our trust doesn't pan out. As long as we start small, trusting people quite broadly is a decision that should pay off in the long run, with the occasional failure a mere cost of doing business. To compensate for when we trust too much, we should consider the costs of failing to trust, the myriad mutually beneficial relationships we could have formed if we had trusted more people.

16

THE CASE AGAINST GULLIBILITY

THIS BOOK IS A LONG argument against the idea that humans are gullible, that they are "wired not to seek truth" and "overly deferential to authority," and that they "cower before uniform opinion."[1] If gullibility appears to have some advantages, allowing us to learn more easily from our elders and our peers, the costs are just too high. The theory of the evolution of communication dictates that for communication to exist, both senders and receivers must benefit from it. If receivers were excessively gullible, they would be mercilessly abused by senders, until they reached a point where they simply stopped paying any attention to what they were being told.

Far from being gullible, we are endowed with a suite of cognitive mechanisms that evaluate what we hear or read. These mechanisms allow us to be open—we listen to information deemed valuable—and vigilant—we reject most harmful messages. As these open vigilance mechanisms grew increasingly complex, we paid attention to more cues telling us that others are right and we are wrong. We let ourselves be influenced by others more and more, going from the fairly limited communicative powers of our predecessors to the infinitely complex and powerful ideas that human language lets us express.

This evolution is reflected in the organization of our minds. People deprived of the most sophisticated means of evaluating information—through brainwashing, subliminal influence, or mere distraction—cannot process the cues telling them to accept new, challenging messages. They revert to a conservative core, rejecting anything they don't already agree with, being much harder, not much easier, to influence.

Open vigilance mechanisms are part of our common cognitive endowment. Their roots can be found in toddlers or even infants. Twelve-month-old infants integrate what they are told with their prior opinions, so that they are easiest to influence when their opinions are weak, and are very stubborn otherwise—as anyone who has interacted with a one-year-old will be painfully aware.[2] Infants this age also track the actions of adults and are more influenced by those who behave competently.[3] Two-and-a-half-year-olds listen more to speakers who offer sound rather than circular arguments.[4] At three years of age, toddlers put more trust in someone who is reporting what they have seen rather than guessed, and they have figured out who is an expert in familiar domain, such as food and toys.[5] When they turn four, preschoolers get a grasp of how best to follow the majority opinion, and they discount agreement based on mere hearsay.[6]

Our open vigilance mechanisms are for learning, and figuring out what to believe and who to trust doesn't stop at four years of age. It never stops: as we accumulate knowledge and experience, we constantly sharpen our open vigilance mechanisms. As an adult, think of how many factors you effortlessly weigh when evaluating the most mundane communication. If your colleague Bao says, "You should switch to the new OS; they've fixed a major security flaw," your reaction will depend on the following: what you already know about the new OS (have you heard it seriously slows computers down?), how vulnerable you think your computer is to attacks (is the security flaw really major?), what Bao's level of competence in this domain is compared with yours (is she the IT specialist?), and whether you believe Bao might have any ulterior motive (might she want you to install the new OS so she can see whether it works well?). None of these kinds of calculations have to be conscious, but they are going on whenever we hear or read something.

In everyday life, when interacting with people we know, cues telling us to change our minds abound: we have time to ascertain goodwill, recognize expertise, and exchange arguments. By contrast, these cues are typically absent from mass persuasion contexts. How can a government agency build trust? How can politicians display their competence to those who don't closely follow politics? How can an advertising campaign convince you a given product is worth buying? Mass persuasion should be tremendously difficult. Indeed, the vast majority of mass persuasion efforts, from propaganda to political campaigns, from religious proselytizing to advertising, end in abject failure. The (modest) successes of mass persuasion are also well accounted for by the functioning of our open vigilance mechanisms. The conclusion reached by Ian Kershaw with respect to Nazi propaganda applies more broadly: the effectiveness of mass persuasion is "heavily dependent on its ability to build on existing consensus, to confirm existing values, to bolster existing prejudices."[7] This reflects the working of plausibility checking, which is always operating, making even the most successful mass persuasion efforts somewhat inert: people might accept the messages, but the messages do not substantially affect their preexisting plans or beliefs. In some situations, when some trust has been built, mass persuasion can change minds, but then only on issues of little personal import, as when people follow political leaders on topics in which they have little interest and even less knowledge.

HOW TO BE WRONG WITHOUT BEING GULLIBLE

If the successes of mass persuasion are, more often than not, a figment of the popular imagination, the dissemination of empirically dubious beliefs is not. We all have, at some point in our lives, endorsed one type of misconception or another, believing in anything from wild rumors about politicians to the dangers of vaccination, conspiracy theories, or a flat earth. Yet the success of these misconceptions is not necessarily a symptom of gullibility.

The spread of most misconceptions is explained by their intuitively appealing content, rather than by the skills of those who propound them. Vaccine hesitancy surfs on the counterintuitiveness of vaccination. Conspiracy theories depend on our justified fear of powerful enemy coalitions. Even flat-earthers argue that you just have to follow your intuition when you look at the horizon and fail to see any curvature.

Even though many misconceptions have an intuitive dimension, most remain cut off from the rest of our cognition: they are reflective beliefs with little consequences for our other thoughts, and limited effects on our actions. The 9/11 truthers might believe the CIA is powerful enough to take down the World Trade Center, but they aren't afraid it could easily silence a blabbing blogger. Most of those who accused Hillary Clinton's aides of pedophilia were content with leaving one-star reviews of the restaurant in which the children were supposedly abused. Even forcefully held religious or scientific beliefs, from god's omniscience to relativity theory, do not deeply affect how we think: Christians still act as if god were an agent who could only pay attention to one thing at a time, and physicists can barely intuit the relationship between time and speed dictated by Einstein's theories.

If some of these reflective beliefs are counterintuitive—an omniscient god, the influence of speed on time—I have argued that most have an intuitive dimension, such as vaccine hesitancy, conspiracy theories, or a flat earth. How can a belief be both reflective (separated from most of our cognition) and intuitive (tapping into a number of our cognitive mechanisms)? Take the belief in a flat earth. Imagine you have no knowledge of astronomy. Someone tells you that the stuff you're standing on, the stuff you see, is called the earth. So far so good. Now they either tell you that the earth is flat, which fits with what you perceive, or that it is spherical, which doesn't. The first alternative is more intuitively compelling. Still, even if you now accept that the earth is flat, the belief remains largely reflective, as you aren't quite sure what to do with the concept of "earth." Unless you're about to embark on a very long journey, or have to perform some astronomical calculations, your ideas about the shape of the earth have no cognitive or practical consequences.

In some cases belief and action, even costly action, go hand in hand: rumors of atrocities committed by the local minority and attacks on that minority, bogus medical theories and harmful medical practices, excessive flattery of a ruler and complete obedience to them. Then, by and large, the beliefs follow the behavior, rather than the other way around. People who want to commit atrocities look for the moral high ground. Doctors like their therapies to be backed up by theories. The political conditions that make it a smart move to obey an autocrat also encourage sycophancy.

Many mistaken but culturally successful ideas serve the interests of those who hold them. People appear more competent by spreading rumors about exaggerated threats. They appear less irrational or immoral by justifying their actions. They credibly communicate their desire to belong to a given group by expressing absurd or odious views that antagonize everybody else. Professing false beliefs doesn't need to be irrational, not by a long shot.

GULLIBLE ABOUT GULLIBILITY?

If people are not gullible, why have scholars and laypeople through the ages, from Plato to Marx, claimed they were? It is often pointed out to me that there seems to be a contradiction between claiming that people are not gullible and saying they wrongly believe others gullible: Isn't the spread of this misconception a sign of gullibility? In fact, the success of the idea that people are gullible can be explained in the same way as other popular mistaken views.

Like most successful rumors, stories about gullibility tend to be false, but intuitively compelling. Those who transmit such stories can score reputation points, as the stories often bear on threats: that words quickly flashed on a cinema screen can control our behavior, that charismatic leaders can turn a tame flock into a bloodthirsty crowd.

Consider the striking crowds in late nineteenth-century France. Out of several hundred demonstrations, the strikers caused only one casualty. Yet it is this unique episode that Émile Zola dramatized in his novel *Germinal*, turning it into the gruesome scene of an angry female crowd castrating their hapless victim. This choice is all the more revealing in that Zola actually sided with the workers. Despite his sympathies, Zola decided to portray a mob running amok in the most sensationalist way: it made a better story than peaceful demonstrators. Ironically, Zola's work later influenced crowd psychologists, who took

Germinal as a faithful description of crowd behavior, using it to condemn the strikers.[8]

As a result of intuitive biases in what we pay attention to, reports of gullibility are likely to become culturally successful, even if they are wildly unrepresentative. It is easier to write a newspaper article about someone who lost their life savings to a Nigerian scam than about the millions of people who laughed at the e-mails. Global rumors—about politicians, celebrities, major events—tend to be false, while local rumors, such as those about our jobs, tend to be accurate. News outlet are, logically enough, only interested in the former (even if only to rebuke them).

As is true of most misconceptions, beliefs in widespread gullibility are largely reflective. Even the most cynical observer, complaining of (what they believe are) gullible citizens voting against their own interests, or gullible consumers buying products they don't want, wouldn't ground their behavior on these beliefs by, say, attempting to talk random strangers into giving them money. The same is true of specific gullibility-related scares. The 1950s panic about subliminal influence didn't stop people from going to the movies. Rumored uses of brainwashing by the New Religious Movements didn't prompt a proportionate legal or popular backlash.

Again, like many other misconceptions, the idea that people are gullible provides post hoc rationalizations for actions or ideas that have other motivations. Until the Enlightenment, accusations of gullibility were routinely used to justify an iniquitous status quo—as it so happens, mostly by people who benefited from that status quo, or who sucked up to those who did. The masses, these scholars claimed, couldn't be trusted with political power, as they would be promptly manipulated by cunning demagogues bent on wrecking the social order. As noted earlier, the perceived danger of demagogues was then "political philosophy's central reason for skepticism about democracy."[9] Supposed widespread gullibility is still recruited today as an argument against democratic power, for example, by Jason Brennan in his book *Against Democracy*.[10]

Ironically, scholars on the other side of the political spectrum, who defend the people's right to a political voice, have also claimed widespread gullibility—not because they feared the population would revolt but because they had to explain why it hadn't already revolted (or, more generally, why it makes the "wrong" political choices). Enlightenment writers who despised the Catholic Church had to explain why its yoke had been docilely borne for centuries (or so they thought). Rousseau tried to absolve the people of any taint, preferring to see the masses as gullible rather than evil: "The people are never corrupted, though often deceived, and it is only then that they seem to will what is evil."[11]

One factor that might explain why a belief in widespread gullibility has proven so successful is quite specific to that belief: the temptation to reverse engineer the tremendous efforts devoted to mass persuasion in our societies. We are inundated with a barrage of advertising, political messages, articles, posts on social media telling us what to drink, eat, buy, feel, think. It can be hard to conceive that such massive efforts don't have a commensurate massive effect on people. However, mass persuasion attempts might be worth the effort even if the audiences are largely skeptical.

Propaganda, for instance, might not convince many people of its content, while still sending a clear signal: that the regime is strong enough to impose its voice. All the way to the top of the one-hundred-foot-tall Column of Trajan spirals a bas-relief depicting the victories of Emperor Trajan in his Eastern European wars. It looks like propaganda aimed at making sure every Roman citizen knew the detail of Trajan's many victories. However, as historian Paul Veyne noted, most of the column's bas-reliefs are simply too high for anyone to see clearly.[12] The message sent by the column isn't what it depicts but its very existence, stating loud and clear that the regime is rich and powerful enough to erect such an edifice.

Or consider a contemporary example. Russian president Vladimir Putin is known to support the ice hockey team SKA Saint Petersburg.[13] Because the team nearly always wins, this could be seen as a propaganda attempt, with Putin basking in the success of his team. However, it is clear to everyone that the SKA Saint Petersburg wins largely because it can break all the rules: the team doesn't respect the salary cap, has its pick of the best players, and is blatantly favored by the referees. The message isn't that Putin is good at picking hockey teams but that he is powerful enough to intimidate everyone into letting his team win, an intrinsically credible message.[14]

Even when mass persuasion bears on the message itself, it can reach its goal without entailing any gullibility on the part of the audience. Many of the products we buy exist in virtually identical versions—different brands of soda, toothpaste, detergent, cigarettes, milk. In these conditions, it is only normal that our minds should respond to minor nudges: the position on the shelf, a tiny discount, or even an appealing ad. Shifts between these essentially identical products may be of no import for consumers, while making a huge difference for companies. Some ads can be cost-effective without any genuine persuasion having taken place.

My Mistake, Your Problem

I have argued that, by and large, popular misconceptions carry little cost for those who hold them, or even serve their social goals. Does this mean, then, that it is not worth trying to refute the false beliefs that spread in spite of our open vigilance mechanisms? That—as a rule—they carry little or no cost for those who hold them doesn't mean they can't be terrible for others.

Before launching the Great Leap Forward, Mao had little understanding of agriculture. He wasn't a farmer relying on his knowledge of plants to feed his family. And so his mechanisms of open vigilance easily went astray, making him accept recommendations simply because they were consistent with his political beliefs. Inspired by the Russian biologist Trofim Lysenko, Mao claimed that plants are like people in the ideal communist state: those of the same class don't compete with each other; instead, "with company they grow easily, when they grow together they will be more comfortable."[15] This led Mao to advocate close cropping: sowing seeds much closer than farmers throughout China had been doing for millennia.

Mao's view on farming had dire consequences—not for him, though, but for the population forced to put them into practice. Close cropping, along with the other counterproductive techniques Mao advocated, led to drastically reduced grain yields. The worst famine of history ensued, killing more than forty million Chinese peasants. Yet Mao stayed in power until his death, in spite of the destruction his asinine ideas had wreaked.

Obviously, the best remedy for this type of disaster is not to improve the critical thinking skills of ruthless dictators, but to get rid of ruthless dictators altogether—or, more generally, to create better feedback loops between decision makers and the effects of their decisions. For all the propaganda and the apparent adulation of Mao, few Chinese peasants would have voted—had they been given the chance—for the guy who had forced close cropping on them. As one of them put it: "We knew about the situation, but no one dared to say anything. If you said anything, they would beat you up. What could we do?"[16]

WE CAN DO BETTER

Turning to democratic societies, I have argued that political elites' influence on public opinion is largely innocuous, as it mostly affects issues on which people have no strong opinion to begin with and are, for them, of little consequence. For instance, in the United States the tradition among political leaders left and right ever since the Cold War had been to take a strongly critical stance of Russia. In his early presidency Donald Trump, a Republican, partly broke with this tradition, to the point that he appeared ready to take the word of Vladimir Putin over that of his own intelligence community.[17] Trump's actions led some Republicans (but no Democrats) to develop a more positive view of Putin, in a typical case of citizens moving toward their favorite political leader's opinion.[18]

For those who developed a more positive opinion of Putin, these views had no personal consequences. Yet this shift could in due course influence policy. The work of political scientist James Stimson (among others) shows that politicians respond to public opinion, being more likely to support policies fitting the popular will.[19] But if the politicians are the ones shaping public opinion in the first place, doesn't this mean they have essentially carte blanche to enact any policy they like, first by creating public support, then by acting on the basis of this support?

Fortunately, a politician's job isn't quite so simple, for two reasons. First, in competitive democracies, citizens heed the opinions of different leaders, who hold different views, pulling the population's opinions in different directions. Second, on most issues a segment of the electorate does have the means and the motivation to have informed opinions. These people don't simply follow whatever their party leader says. Instead, they form their own opinion on the basis of their personal experiences, what they see in the news, what they read in newspapers, and so forth. The views of these informed citizens are the signal in the noise of public opinion, and they largely guide its movements.[20]

Trump has been famously tough on immigration. However, far from following his lead, members of his party have developed more lenient views on the topic. Compared with 2015, in 2018 Republicans were less likely to want legal immigration to be decreased.[21] These shifts are explained by what political scientists call the *thermostatic model* of public opinion. When politicians veer too far in a given direction, people who pay attention express their disagreement by moving their opinions the other way.[22]

It is quite plausible that the movements in public opinion influencing policy are largely shaped by informed individuals who don't simply follow their party's lead. The lack of effort by other citizens, then, is not disastrous, but it still reveals some tantalizing possibilities. If more voters were as reactive as the informed minority to actual events, shifts in public opinion would be faster and stronger and would have more impact on policy. But mass persuasion is hard, for better or worse. Getting people who know and care very little about politics to abandon the easy strategy of following their party's lead is not easy.

FRAGILE CHAINS OF TRUST

The take-home message of this book is: influencing people isn't too easy, but too hard. Most of the misconceptions we have explored persist because people refuse to believe those who know better. False rumors and conspiracy theories survive long after they have been debunked. Quack doctors and flat-earthers ignore all the scientific evidence thrown at them.

Take anti-vaxxers. If we assume an intuitive reaction against vaccination, the issue with anti-vaxxers isn't that they are not vigilant enough but that they are not open enough. People who have access to the relevant medical information, and who still refuse basic vaccines, are failing to put their trust in the right place—medical professionals, the scientific consensus—and to be convinced by sound arguments.[23] This is what we must work on. Pharmaceutical companies engage in a variety of practices that provide grounds for mistrust, from failing to report unsuccessful clinical trials to buying doctors' influence.[24] Cleaning up their acts would help allay some of the mistrust. It is also important to properly engage with anti-vaxxers. Unfortunately, most people have only limited argumentative tools to do so.[25] Those who find the right arguments, such as experts who take the time to engage vaccine doubters in conversation, are more likely to be convincing.[26]

The same logic applies to other domains. As many find conspiracy theories intuitively compelling, attempts to shut off the channels through which conspiracy theories spread cannot eradicate them. Even the Chinese regime's tight control over the media doesn't stop conspiracy theories from flourishing.[27] The best way to curb the spread of conspiracy theories, surely, is to have a trustworthy government with strong laws against corruption, conflicts of interests, regulatory capture, and so forth.[28]

This is, presumably, why conspiracy theories are less prevalent in Norway than in Pakistan.[29]

The example of science shows how the foundations of an institution affect its public perception. Nearly all scientific theories are deeply counterintuitive, and yet they have percolated through most layers of society. This has happened even though few people know scientists personally, and even fewer have a genuine understanding of the arguments supporting, say, relativity theory or evolution by natural selection. The wide spread of scientific ideas, in spite of their apparent implausibility, has been underpinned by the solid, even if not unblemished, foundations of trust the scientific enterprise rests on.

These foundations of trust must be protected and buttressed. As an example, my own discipline of psychology is addressing a number of long-standing problems, as we strive to improve our statistical practices, recruit more diverse samples of participants, reduce conflicts of interest, run experiments several times to have more faith in their results, and commit to our hypotheses ahead of conducting a study, to avoid post hoc interpretations. Other disciplines from medicine to economics are tackling similar issues. They will come out stronger from this "credibility revolution," and, in time, the improved trustworthiness will affect the spread of scientific advances throughout society.

We aren't gullible: by default we veer on the side of being resistant to new ideas. In the absence of the right cues, we reject messages that don't fit with our preconceived views or preexisting plans. To persuade us otherwise takes long-established, carefully maintained trust, clearly demonstrated expertise, and sound arguments. Science, the media, and other institutions that spread accurate but often counterintuitive messages face an uphill battle, as they must transmit these messages and keep them credible along great chains of trust and argumentation. Quasi-miraculously, these chains connect us to the latest scientific discoveries and to events on the other side of the planet. We can only hope for new means of strengthening and extending these ever-fragile links.

NOTES

INTRODUCTION

1. Mark Sargant, prominent flat-earther, in the documentary *Behind the Curve*.
2. Although they are elsewhere called *epistemic vigilance*; see Sperber et al., 2010.

CHAPTER 1

1. Dickson, 2007, p. 10.
2. Thucydides, *The history of the Peloponnesian War*, http://classics.mit.edu/Thucydides/pelopwar.mb.txt (accessed July 19, 2018).
3. Plato, *Republic*, Book VIII, 565a, trans. Jowett; see also 488d. http://classics.mit.edu/Plato/republic.9.viii.html (accessed July 19, 2018).
4. Some scholars (e.g., Greenspan, 2008) have attempted to differentiate *credulity* ("this term refers to a tendency to believe things that on their face are ridiculous or that lack adequate supporting evidence") and *gullibility* ("the term gullibility really refers to a pattern of being duped, which repeats itself in different settings, even in the face of warning signs"). Here I will use the terms largely interchangeably, to refer to when people are influenced by what others tell them without having any good reason for being so influenced.
5. Holbach, 1835, p. 119.
6. "La crédulité des premières dupes" (Condorcet, 1797, p. 22); "charlatans et leurs sorciers" (p. 21). See similar references in Singh, 2018.
7. Peires, 1989, location 2060–2062.
8. Eric Killelea, "Alex Jones' mis-infowars: 7 Bat-Sh*t Conspiracy Theories," *Rolling Stone*, February 21, 2017, http://www.rollingstone.com/culture/lists/alex-jones-mis-infowars-7-bat-sht-conspiracy-theories-w467509/the-government-is-complicit-in-countless-terrorist-and-lone-gunman-attacks-w467737.

NOTES TO CHAPTER 1

9. Callum Borchers, "A harsh truth about fake news: Some people are super gullible," *Washington Post*, December 5, 2016, https://www.washingtonpost.com/news/the-fix/wp/2016/11/22/a-harsh-truth-about-fake-news-some-people-are-super-gullible/.
10. Heckewelder, 1876, p. 297.
11. Dawkins, 2010, p. 141.
12. Truther monkey (@Thedyer1971), "The mind controlled sheeple. Welcome to the new world order," Twitter, September 26, 2017, 12:53 a.m., https://twitter.com/Thedyer1971/status/912585964978966528.
13. Borchers, "A Harsh truth about fake news"; more generally, see Donovan, 2004, which shows how often accusations of gullibility are hurled from each side.
14. Marcuse, 1966, pp. 46, 15; see also Abercrombie, Hill, & Turner, 1980. For a much more nuanced take on the role of the dominant ideology, see the work of Antonio Gramsci (for an introduction, see Hoare & Sperber, 2015).
15. Stanley, 2015, p. 27.
16. Paul Waldman, "Trump and republicans treat their voters like morons," *Washington Post*, July 26, 2017, https://www.washingtonpost.com/blogs/plum-line/wp/2017/07/26/trump-and-republicans-treat-their-voters-like-morons/.
17. Asch, 1956. Source for figure 1: https://en.wikipedia.org/wiki/Asch_conformity_experiments#/media/File:Asch_experiment.svg (accessed November 21, 2018), CC BY-SA 4.0.
18. Moscovici, 1985, p. 349, cited by Friend, Rafferty, & Bramel, 1990.
19. Milgram, Bickman, & Berkowitz, 1969.
20. Milgram, 1974.
21. Perry, 2013, location 145.
22. Brennan, 2012, p. 8.
23. Gilbert, Krull, & Malone, 1990, p. 612.
24. Heraclitus, 2001, fragment 111.
25. David Robson, "Why are people so incredibly gullible?," BBC, March 24, 2016, http://www.bbc.com/future/story/20160323-why-are-people-so-incredibly-gullible.
26. This was actually not the case for Asch; see Friend et al., 1990.
27. Hirschfeld, 2002. While this may be true of the majority of anthropologists, who did not pay much attention to children, there is also in anthropology and social psychology an old and strong tradition of work on acculturation (e.g., Linton, 1963).
28. Boyer, 1994, p. 22.
29. Strauss & Quinn, 1997, p. 23.
30. Dawkins, 2010, p. 134.
31. See Henrich, 2015.

32. Boyd & Richerson, 1985; Richerson & Boyd, 2005. Their work focuses mostly on material culture, for which the issue of gullibility is possibly less pressing than in the case of communication.
33. E.g., Barkow, Cosmides, & Tooby, 1992; Pinker, 1997.
34. Henrich, 2015.
35. On Laland's social learning strategies, see, e.g., Laland, 2004.
36. Or conformist transmission, a member of a family of "frequency-based biases"; see Boyd & Richerson, 1985; Henrich & Boyd, 1998. For a recent criticism of the usefulness of these strategies, see Grove, 2018.
37. Henrich & Gil-White, 2001; for a review of recent evidence, see Jiménez & Mesoudi, 2019.
38. K. Hill & Kintigh, 2009.
39. Richerson & Boyd, 2005, pp. 162–167, 187.
40. Boyd & Richerson, 1985, pp. 204ff; see also Nunn & Sanchez de la Sierra, 2017 (for a critique of this last reference, see Lou Keep, "The use and abuse of witchdoctors for life," *Samzdat*, June 19, 2017, https://samzdat.com/2017/06/19/the-use-and-abuse-of-witchdoctors-for-life/.
41. Henrich, 2015, p. 49.
42. Richerson & Boyd, 2005, p. 124.
43. Boyd & Richerson, 2005.
44. Boyd & Richerson, 2005, p. 18.
45. Marx & Engels, 1970, p. 64.

CHAPTER 2

1. Caro, 1986a.
2. Ostreiher & Heifetz, 2017; Sommer, 2011.
3. See references in Haig, 1993.
4. Wray, Klein, Mattila, & Seeley, 2008, which suggests that prior contrary results (Gould & Gould, 1982) were artifacts. See also Dunlap, Nielsen, Dornhaus, & Papaj, 2016.
5. Scott-Phillips, 2008, 2014; Scott-Phillips, Blythe, Gardner, & West, 2012.
6. Seyfarth, Cheney, & Marler, 1980.
7. Nishida et al., 2006.
8. Dawkins & Krebs, 1978; Krebs & Dawkins, 1984; Maynard Smith & Harper, 2003.
9. Haig, 1993, 1996.
10. Haig, 1993, p. 511.
11. Blumstein, Steinmetz, Armitage, & Daniel, 1997. For vervet monkeys, one of the mechanisms that allows the alarm calls to remain stable is that vervets can learn to ignore

the signals of unreliable callers, thus keeping the costs of dishonest signals low and providing an incentive for callers to send honest signals (Cheney & Seyfarth, 1988).

12. J. Wright, 1997; J. Wright, Parker, & Lundy, 1999.

13. See C. T. Bergstrom & Lachmann, 2001.

14. O. Hasson, 1991.

15. Caro, 1986b.

16. For all the evidence that follows, see Caro, 1986a, 1986b; FitzGibbon & Fanshawe, 1988.

17. Nelissen & Meijers, 2011.

18. E.g., Henrich, 2009; Iannaccone, 1992.

19. E. A. Smith & Bird, 2000.

20. See, e.g., Higham, 2013.

21. Borgia, 1985. For more references, see Madden, 2002.

22. Zahavi & Zahavi, 1997.

23. Borgia, 1993.

24. Madden, 2002.

CHAPTER 3

1. Dubreuil, 2010; Sterelny, 2012.

2. Dediu & Levinson, 2018; Hoffmann et al., 2018; see also Andrew Lawler, "Neandertals, Stone Age people may have voyaged the Mediterranean," *Science*, April 24, 2018, http://www.sciencemag.org/news/2018/04/neandertals-stone-age-people-may-have-voyaged-mediterranean.

3. Dan Sperber and his colleagues made this point in a 2010 article: Sperber et al., 2010; see also Clément, 2006; Harris, 2012; O. Morin, 2016.

4. Cited in Carruthers, 2009, p. 175.

5. Anthony, 1999.

6. Cited in Carruthers, 2009, p. 192.

7. *Life* magazine, cited in Carruthers, 2009, p. 192.

8. Pratkanis & Aronson, 1992, chap. 34.

9. Pratkanis & Aronson, 1992, chap. 34.

10. Reicher, 1996.

11. Cited in Barrows, 1981, p. 48.

12. Cited in Barrows, 1981, p. 47.

13. Taine, 1876, p. 226.

14. F. G. Robinson, 1988, p. 387.

15. Paul Waldman, "Trump and Republicans treat their voters like morons," *Washington Post*, July 26, 2017, https://www.washingtonpost.com/blogs/plum-line/wp/2017/07/26/trump-and-republicans-treat-their-voters-like-morons/; Jason Brennan,

"Trump won because voters are ignorant, literally," *Foreign Policy*, November 10, 2016, http://foreignpolicy.com/2016/11/10/the-dance-of-the-dunces-trump-clinton-election-republican-democrat/.

16. Peter Kate Piercy, "Classist innuendo about educated Remain voters and the 'white man van' of Leave has revealed something very distasteful about Britain," *Independent*, June 20, 2016, http://www.independent.co.uk/voices/classist-innuendo-about-educated-remain-voters-and-the-white-van-men-of-leave-has-revealed-something-a7091206.html.

17. Zimbardo, Johnson, & McCann, 2012, p. 286.
18. Myers, 2009, p. 263.
19. Bonnefon, Hopfensitz, & De Neys, 2017; Todorov, Funk, & Olivola, 2015.
20. There are many issues with these dual-process models, which Dan Sperber and I have previously criticized; see Mercier & Sperber, 2017.
21. Frederick, 2005.
22. Although most people only do so when the correct answer is explained by someone else. Most people who provide the correct answer on their own do so intuitively; see Bago & De Neys, 2019.
23. Gilbert et al., 1990; Gilbert, Tafarodi, & Malone, 1993.
24. Gilbert et al., 1993.
25. Kahneman, 2011, p. 81.
26. Gervais & Norenzayan, 2012.
27. Aarnio & Lindeman, 2005; Pennycook, Cheyne, Seli, Koehler, & Fugelsang, 2012.
28. Tyndale-Biscoe, 2005, p. 234.
29. Ratcliffe, Fenton, & Galef, 2003.
30. Rozin, 1976, p. 28.
31. Rozin, 1976.
32. Garcia, Kimeldorf, & Koelling, 1955.
33. Rozin, 1976, p. 26.
34. Rozin, 1976.
35. Cheney & Seyfarth, 1990.
36. de Waal, 1982.
37. Cheney, Seyfarth, & Silk, 1995.
38. Desrochers, Morissette, & Ricard, 1995.
39. Tomasello, Call, & Gluckman, 1997.
40. J. Wood, Glynn, Phillips, & Hauser, 2007.
41. An argument I had previously made in Mercier, 2013.
42. Carruthers, 2009.
43. Alexander & Bruning, 2008; Meissner, Surmon-Böhr, Oleszkiewicz, & Alison, 2017.

44. Pratkanis & Aronson, 1992.
45. Pratkanis & Aronson, 1992; see also Trappey, 1996.
46. Strahan, Spencer, & Zanna, 2002. Experiments on subliminal influence might not be fully reliable, as many of them are embroiled in a replication controversy: even if the results were obtained once, it is not clear they can be obtained again, so that the initial results might have been a statistical fluke (e.g., Open Science Collaboration, 2015).
47. Richter, Schroeder, & Wöhrmann, 2009.
48. U. Hasson, Simmons, & Todorov, 2005.
49. Kahneman, 2011, p. 81.
50. B. Bergstrom & Boyer, submitted. For more on these debates, see Isberner & Richter, 2013, 2014; Sklar et al., 2012; Wiswede, Koranyi, Müller, Langner, & Rothermund, 2012.
51. Gervais et al., 2018.
52. Majima, 2015.
53. Mascaro & Morin, 2014.
54. Couillard & Woodward, 1999.
55. Mascaro & Morin, 2014.

CHAPTER 4

1. Nyhan & Reifler, 2010.
2. Nyhan & Reifler, 2015.
3. Bonaccio & Dalal, 2006; Yaniv, 2004; the one-third figure can be found, for instance, in Yaniv & Kleinberger, 2000. This is actually slightly misleading. In fact, on each item, two-thirds of the people don't change their minds, and one-third adopt the other opinion wholesale (as a first approximation). People's preference for choosing one of the two opinions (theirs or that of the other individual), instead of averaging between the two, is suboptimal (Larrick & Soll, 2006).
4. T. Wood & Porter, 2016.
5. Aird, Ecker, Swire, Berinsky, & Lewandowsky, 2018; Chan, Jones, Hall Jamieson, & Albarracin, 2017; De Vries, Hobolt, & Tilley, 2018; Dixon, Hmielowski, & Ma, 2017; Dockendorff & Mercier, in preparation; Ecker, O'Reilly, Reid, & Chang, 2019; Facchini, Margalit, & Nakata, 2016; Grigorieff, Roth, & Ubfal, 2018; Guess & Coppock, 2015, 2018; S. J. Hill, 2017; Hopkins, Sides, & Citrin, 2019; J. W. Kim, 2018; Leeper & Slothuus, 2015; Nair, 2018; Nyhan, Porter, Reifler, & Wood, 2017; Tappin & Gadsby, 2019; van der Linden, Maibach, & Leiserowitz, 2019; Walter & Murphy, 2018.
6. For some evidence that people are not good at doing this when the task is very explicit, see Dewitt, Lagnado, & Fenton, submitted.

7. See Thagard, 2005.
8. I thank Jennifer Nagel for introducing it to me.
9. Trouche, Sander, & Mercier, 2014.
10. Claidière, Trouche, & Mercier, 2017.
11. Mercier, 2012; Mercier, Bonnier, & Trouche, 2016; Mercier & Sperber, 2011, 2017.
12. Sperber & Mercier, 2018.
13. Plato, *Meno*, Jowett translation, https://en.wikisource.org/wiki/Meno (accessed on July 28 2019).
14. There are exceptions: people might want to justify their views (to display their rationality) or to limit their exposure if what they say turns out to be mistaken; see Mercier & Sperber, 2017.
15. Liberman, Minson, Bryan, & Ross, 2012; Minson, Liberman, & Ross, 2011.
16. See references in Mercier, 2016a.
17. Trouche, Shao, & Mercier, 2019; for evidence in children, see Castelain, Bernard, Van der Henst, & Mercier, 2016.
18. For reviews, see Hahn & Oaksford, 2007; Petty & Wegener, 1998; see also, for some observational evidence, Priniski & Horne, 2018. Other experiments appear to show that we are biased when we evaluate arguments that challenge our beliefs (e.g., Edwards & Smith, 1996; Greenwald, 1968; Taber & Lodge, 2006). However, it can be argued that this apparent bias stems from the production of arguments, rather than the evaluation of arguments, suggesting that the evaluation of arguments is in fact unbiased (Mercier, 2016b; Trouche et al., 2019).
19. "Incompleteness theorems," Wikipedia, https://en.wikipedia.org/wiki/G%C3%B6del%27s_incompleteness_theorems (accessed April 24, 2019).
20. Mancosu, 1999.
21. Planck, 1968, pp. 33–34.
22. Nitecki, Lemke, Pullman, & Johnson, 1978; Oreskes, 1988. For other examples, see Cohen, 1985; Kitcher, 1993; Wootton, 2015.
23. Mercier & Sperber, 2017, chap. 17.
24. Mansbridge, 1999.
25. Cited, translated, and discussed in Galler, 2007.
26. Shtulman, 2006; Shtulman & Valcarcel, 2012.
27. Miton & Mercier, 2015.
28. Durbach, 2000, p. 52.
29. Elena Conis, "Vaccination Resistance in Historical Perspective," The American Historian, http://tah.oah.org/issue-5/vaccination-resistance/ (accessed July 17, 2018).
30. M. J. Smith, Ellenberg, Bell, & Rubin, 2008.
31. Boyer & Petersen, 2012, 2018; van Prooijen & Van Vugt, 2018.

CHAPTER 5

1. "'Je devais aller à Bruxelles, je me suis retrouvée à Zagreb': l'incroyable périple en auto de Sabine, d'Erquelinnes," Sudinfo, January 11, 2013, https://www.sudinfo.be/art/640639/article/regions/charleroi/actualite/2013-01-11/%C2%ABje-devais-aller-a-bruxelles-je-me-suis-retrouvee-a-zagreb%C2%BB-l-incroyable-p.

2. E. J. Robinson, Champion, & Mitchell, 1999. For reviews of the enormous amount of work done on how children evaluate testimony, see Clément, 2010; Harris, 2012; Harris, Koenig, Corriveau, & Jaswal, 2018.

3. Castelain, Girotto, Jamet, & Mercier, 2016; Mercier, Bernard, & Clément, 2014; Mercier, Sudo, Castelain, Bernard, & Matsui, 2018.

4. On how people who tell us things we agree with are deemed more trustworthy, see Collins, Hahn, von Gerber, & Olsson, 2018.

5. Choleris, Guo, Liu, Mainardi, & Valsecchi, 1997.

6. See Analytis, Barkoczi, & Herzog, 2018.

7. Malkiel & McCue, 1985; Taleb, 2005.

8. K. Hill & Kintigh, 2009.

9. Or fishing in small-scale societies, see Henrich & Broesch, 2011.

10. Howard, 1983; Sternberg, 1985. Neither is supported by much data.

11. This position is bolstered, for instance, by the weakness of transfer effects in learning, a phenomenon already recognized by Thorndyke in the early twentieth century: "The mind is so specialized into a multitude of independent capacities that we alter human nature only in small spots" (1917, p. 246). For recent references, see Sala et al., 2018; Sala & Gobet, 2017, 2018.

12. Kushnir, Vredenburgh, & Schneider, 2013; VanderBorght & Jaswal, 2009.

13. Keil, Stein, Webb, Billings, & Rozenblit, 2008; Lutz & Keil, 2002.

14. Stibbard-Hawkes, Attenborough, & Marlowe, 2018.

15. Brand & Mesoudi, 2018.

16. Huckfeldt, 2001, see also Katz & Lazarsfeld, 1955; Lazarsfeld, Berelson, & Gaudet, 1948.

17. Kierkegaard, 1961, p. 106.

18. Mark Twain, *The complete works of Mark Twain*, p. 392, Archive.org, https://archive.org/stream/completeworksofm22twai/completeworksofm22twai_djvu.txt (accessed July 19, 2018).

19. Mercier, Dockendorff, & Schwartzberg, submitted.

20. Condorcet, 1785.

21. Galton, 1907; see also Larrick & Soll, 2006. Galton actually used the median, not the mean, because computing the mean would have been impossible by hand (thanks to Emile Servan-Schreiber for pointing this out).

NOTES TO CHAPTER 6   281

22. Surowiecki, 2005.
23. Source for figure 2: https://xkcd.com/1170/ (accessed June 24, 2019).
24. Conradt & List, 2009; Conradt & Roper, 2003.
25. Strandburg-Peshkin, Farine, Couzin, & Crofoot, 2015.
26. Hastie & Kameda, 2005.
27. T.J.H. Morgan, Rendell, Ehn, Hoppitt, & Laland, 2012.
28. Mercier & Morin, 2019.
29. Mercier & Morin, 2019.
30. Dehaene, 1999.
31. For a review, see Mercier & Morin, 2019.
32. E.g., Maines, 1990.
33. Mercier & Miton, 2019.
34. J. Hu, Whalen, Buchsbaum, Griffiths, & Xu, 2015; but see Einav, 2017.
35. For a review, see Mercier & Morin, 2019.
36. See Friend et al., 1990; Griggs, 2015.
37. On the classic informational/normative conformity distinction, see Deutsch & Gerard, 1955; for difficulties with it, see Hodges & Geyer, 2006. For similar results with children, see Corriveau & Harris, 2010; Haun & Tomasello, 2011.
38. Allen, 1965, p. 143. Allen also reports experiments showing that people revert to the correct answer when they are asked again, in the absence of the group.
39. Asch, 1956, p. 56.
40. Asch, 1956, p. 47.
41. Gallup, Chong, & Couzin, 2012; Gallup, Hale, et al., 2012.
42. Clément, Koenig, & Harris, 2004.
43. On competence, see Bernard, Proust, & Clément, 2015.
44. A great description of fool's errands is provided by Umbres, 2018.

CHAPTER 6

1. DePaulo et al., 2003.
2. Freud, 1905, p. 94; cited in Bond, Howard, Hutchison, & Masip, 2013.
3. Qing China: Conner, 2000, p. 142; ancient India: Rocher, 1964, p. 346; European Middle Ages: Ullmann, 1946; Robisheaux, 2009, p. 206; twentieth-century United States: Underwood, 1995, pp. 622ff.
4. Kassin & Gudjonsson, 2004, citing the textbook by Inbau, Reid, Buckley, & Jayne, 2001.
5. E.g., Ekman, 2001, 2009.
6. Weinberger, 2010. For direct evidence that micro-expressions training doesn't work, see Jordan et al., in press.

7. Porter & ten Brinke, 2008.
8. ten Brinke, MacDonald, Porter, & O'Connor, 2012.
9. DePaulo, 1992.
10. DePaulo et al., 2003; Hartwig & Bond, 2011; Vrij, 2000.
11. Hartwig & Bond, 2011.
12. Honts & Hartwig, 2014, p. 40; see also Foerster, Wirth, Herbort, Kunde, & Pfister, 2017; Luke, in press; Raskin, Honts, & Kircher, 2013; Roulin & Ternes, 2019. More recently, it has been shown that people cannot distinguish between actual emotional screams and acted screams; Engelberg & Gouzoules, 2019.
13. For a review, see Bond & DePaulo, 2006; see also Bond, 2008.
14. Levine, 2014; see also Gilbert, 1991.
15. E.g., DePaulo, Kashy, Kirkendol, Wyer, & Epstein, 1996. In other cultures, lying is much more frequent, and, as a result, people are more attuned to the possibility of lying (e.g., Gilsenan, 1976).
16. Reid, 1970, chap. 24.
17. See, e.g., Helen Klein Murillo, "The law of lying: Perjury, false statements, and obstruction," Lawfare, March 22, 2017, https://www.lawfareblog.com/law-lying-perjury-false-statements-and-obstruction.
18. To some extent, this means that there shouldn't be strong pressures for people to self-deceive (contra Simler & Hanson, 2017; von Hippel & Trivers, 2011). If bad information is punished about equally whether it was sent intentionally or not, there is no need to self-deceive; see Mercier, 2011.
19. E.g., Birch & Bloom, 2007.
20. From the point of view of open vigilance, it is also worth noting that however diligent we expect someone to be, at best this means their opinion will be deemed as valuable as ours. It is only if we have other reasons to believe them—they are more competent, they have good reasons—that we should go more than 50 percent of the way toward what they tell us.
21. Sniezek, Schrah, & Dalal, 2004.
22. Gino, 2008.
23. Gendelman, 2013.
24. In fact, the alignment of incentives probably played a crucial role only at the beginning, while the mechanism I describe next takes over later as the main driver of their cooperation and continued friendship.
25. Reyes-Jaquez & Echols, 2015.
26. This is likely why chimpanzees ignore pointing altogether: they assume that the individual pointing is not cooperating but competing with them.
27. Mills & Keil, 2005; see also Mills & Keil, 2008.
28. Meissner & Kassin, 2002; Street & Richardson, 2015.

29. The problem is compounded by the fact that knowing what other people's incentives are is difficult. We are better off erring on the side of caution, believing, in the absence of clear contrary evidence, their incentives not to be aligned with ours, making communication even less likely.

30. Frank, 1988.

31. Technically, the opinion each potential interlocutor has of them. This might be different from their reputation, which is a commonly agreed upon opinion toward the speaker; Sperber & Baumard, 2012.

32. Boehm, 1999.

33. Baumard, André, & Sperber, 2013.

34. For a related point, see Shea et al., 2014.

35. Van Zant & Andrade, submitted.

36. Brosseau-Liard & Poulin-Dubois, 2014; see also, e.g., Matsui, Rakoczy, Miura, & Tomasello, 2009. For results on adults, see Bahrami et al., 2010; Fusaroli et al., 2012; Pulford, Colman, Buabang, & Krockow, 2018.

37. Tenney, MacCoun, Spellman, & Hastie, 2007; Tenney et al., 2008, 2011. Some studies suggest that people do not adjust their trust in others as a function of past overconfidence enough (Anderson, Brion, Moore, & Kennedy, 2012; J. A. Kennedy, Anderson, & Moore, 2013). However, even these studies show a drop in trust in overconfident speakers, and this drop would likely increase if speakers kept on being overconfident (see Vullioud, Clément, Scott-Phillips, & Mercier, 2017).

38. Vullioud et al., 2017. We also show that if people trust a speaker for reasons other than the source expressing confidence, then they drop their trust in the speaker less once the speaker is proven to have been mistaken.

39. Boehm, 1999; Chagnon, 1992; Safra, Baumard, & Chevallier, submitted.

40. E.g., Kam & Zechmeister, 2013; Keller & Lehmann, 2006.

41. Amos, Holmes, & Strutton, 2008.

42. Laustsen & Bor, 2017.

43. Amos et al., 2008.

44. Knittel & Stango, 2009.

45. On promises: e.g., Artés, 2013; Pomper & Lederman, 1980; Royed, 1996; on corruption: Costas-Pérez, Solé-Ollé, & Sorribas-Navarro, 2012.

CHAPTER 7

1. Rankin & Philip, 1963, p. 167. What follows about Tanganyika is mostly based on this paper; see also Ebrahim, 1968.

2. See, e.g., Evans & Bartholomew, 2009.

3. Susan Dominus, "What happened to the girls in Le Roy," *New York Times Magazine*, March 7, 2012, https://www.nytimes.com/2012/03/11/magazine/teenage-girls-twitching-le-roy.html.

4. Rankin & Philip, 1963, p. 167.

5. Le Bon, 1897.

6. Tarde, 1892, p. 373 (my translation).

7. Sighele, 1901, p. 48 (my translation).

8. See Warren & Power, 2015.

9. A. Brad Schwartz, "Orson Welles and history's first viral-media event," *Vanity Fair*, April 27, 2015, https://www.vanityfair.com/culture/2015/04/broadcast-hysteria-orson-welles-war-of-the-worlds.

10. Moorehead, 1965, p. 226.

11. Coviello et al., 2014; Kramer, Guillory, & Hancock, 2014.

12. Canetti, 1981, p. 77.

13. Sighele, 1901, p. 190 (my translation).

14. Le Bon, 1900, p. 21 (my translation).

15. Lanzetta & Englis, 1989.

16. Dimberg, Thunberg, & Elmehed, 2000.

17. Dezecache et al., submitted.

18. Hatfield, Cacioppo, & Rapson, 1994, p. 5.

19. Cited in Sighele, 1901, p. 59 (my translation); for a more recent reference, see Moscovici, 1981.

20. Frank, 1988; Sell, Tooby, & Cosmides, 2009.

21. Burgess, 1839, p. 49.

22. Frank, 1988.

23. Indeed, this strategy seems to work; see Reed, DeScioli, & Pinker, 2014.

24. Frank, 1988, p. 121; see also Owren & Bachorowski, 2001.

25. Fodor, 1983.

26. More things are under conscious control than we often believe. Good actors can manipulate their facial expressions at will. Some people can even consciously affect piloerection (Heathers, Fayn, Silvia, Tiliopoulos, & Goodwin, 2018).

27. Dezecache, Mercier, & Scott-Phillips, 2013.

28. Tamis-LeMonda et al., 2008; see also G. Kim & Kwak, 2011.

29. Chiarella & Poulin-Dubois, 2013; see also Chiarella & Poulin-Dubois 2015.

30. Hepach, Vaish, & Tomasello, 2013.

31. Lanzetta & Englis, 1989.

32. Zeifman & Brown, 2011.

33. Hofman, Bos, Schutter, & van Honk, 2012.

34. Weisbuch & Ambady, 2008; see also Han, 2018.

35. For more references, see Dezecache et al., 2013; Norscia & Palagi, 2011.
36. Campagna, Mislin, Kong, & Bottom, 2016.
37. On how poor the analogy is, see Warren & Power, 2015.
38. Crivelli & Fridlund, 2018.
39. McPhail, 1991; see O. Morin, 2016.
40. A similar point is made by "Beyond contagion: Social identity processes in involuntary social influence," *Crowds and Identities: John Drury's Research Group*, University of Sussex, http://www.sussex.ac.uk/psychology/crowdsidentities/projects/beyondcontagion (accessed July 20, 2018).
41. This section relies on Boss, 1997.
42. Dominus, "What happened to the girls in Le Roy."
43. Evans & Bartholomew, 2009, see also Ong, 1987; Boss, 1997, p. 237.
44. Lopez-Ibor, Soria, Canas, & Rodriguez-Gamazo, 1985, p. 358.
45. Couch, 1968; Drury, Novelli, & Stott, 2013; McPhail, 1991; Schweingruber & Wohlstein, 2005.
46. Taine, 1885 book 1, chap. V.
47. Rudé, 1959.
48. Barrows, 1981.
49. Barrows, 1981.
50. E.g., J. Barker, 2014; more generally, see Hernon, 2006.
51. Cited in White, 2016.
52. Klarman, 2016.
53. Wang, 1995, p. 72.
54. Taine, 1876, p. 241.
55. McPhail, 1991, in particular pp. 44ff.; Tilly & Tilly, 1975.
56. If, moreover, there is a gradation in how many people already behaving badly it takes to allow someone to behave badly, then the phenomenon looks like a cascade of influence, when in fact no (direct) influence takes place at all; see Granovetter, 1978.
57. Here, I rely on Dezecache, 2015, and Mawson, 2012, in particular pp. 234ff.
58. Jefferson Pooley and Michael J. Sokolow, "The myth of the *War of the Worlds* panic," October 28, 2013, *Slate*, http://www.slate.com/articles/arts/history/2013/10/orson_welles_war_of_the_worlds_panic_myth_the_infamous_radio_broadcast_did.html. See also Lagrange, 2005.
59. Janis, 1951.
60. Schultz, 1964.
61. Proulx, Fahy, & Walker, 2004.
62. Dezecache et al., submitted.
63. Dezecache et al., submitted; see also Johnson, 1988.
64. R. H. Turner & Killian, 1972.

65. McPhail, 2007.
66. Aveni, 1977; Johnson, Feinberg, & Johnston, 1994; McPhail & Wohlstein, 1983.
67. See references in Mawson, 2012, pp. 143ff.

CHAPTER 8

1. This chapter and the next draw on Mercier, 2017. The "worst enemies" is from Signer, 2009.
2. Signer, 2009, pp. 40–41.
3. Thucydides, *The History of the Peloponnesian War*, http://classics.mit.edu/Thucydides/pelopwar.mb.txt (accessed November 23, 2018).
4. See "Mytilenean revolt," Wikipedia, https://en.wikipedia.org/wiki/Mytilenean_revolt (accessed November 23, 2018).
5. Thucydides, *The History of the Peloponnesian War*, 3.37, http://classics.mit.edu/Thucydides/pelopwar.mb.txt (accessed November 23, 2018).
6. *Republic*, Book VIII, 565a, trans. Jowett; see also 488d, http://classics.mit.edu/Plato/republic.9.viii.html (accessed November 23, 2018).
7. "Cleon," in William Smith (Ed.), *A dictionary of Greek and Roman biography and mythology*, http://www.perseus.tufts.edu/hopper/text?doc=Perseus:text:1999.04.0104:entry=cleon-bio-1 (accessed November 23, 2018).
8. "Cleon."
9. Whedbee, 2004.
10. See in particular Kershaw, 1983b.
11. Kershaw, 1987; see also Kershaw, 1983b, 1991.
12. Kershaw, 1987, p. 46.
13. Kershaw, 1987, p. 46.
14. Selb & Munzert, 2018, p. 1050
15. Kershaw, 1987, pp. 61, 123; see also Voigtländer & Voth, 2014.
16. Kershaw, 1987, p. 146.
17. Kershaw, 1987, pp. 187–188.
18. Kershaw, 1987, pp. 194ff.
19. Kershaw, 1987, p. 147.
20. Kershaw, 1987, pp. 233ff.
21. See, e.g., Wang, 1995, for Mao.
22. This analogy is developed in Watts, 2011, pp. 96–97.
23. This account is drawn from Peires, 1989.
24. Peires, 1989, location 2060–2062.
25. Peires, 1989, location 363–365.
26. Peires, 1989, locations 1965–1966, 1921–1923.
27. Peires, 1989, location 1923–1927.

28. Peires, 1989, location 4257–4262.
29. Peires, 1989, location 4262–4264.
30. Peires, 1989, locations 2550–2552, 2078–2081.
31. Peires, 1989, location 3653–3657.
32. Peires, 1989, locations 2524–2526, 3672–3676.
33. Peires, 1989, locations 3699–3700, 4369–4370.
34. Peires, 1989, location 46–49.
35. Stapleton, 1991.
36. Peires, 1989, location 4483–4486.
37. Stapleton, 1991, p. 385.
38. Peires, 1989, location 4577–4582.
39. Cohn, 1970, p. 66; for the "children's crusade," see Dickson, 2007.
40. Weber, 2000; see also Barkun, 1986; Hall, 2013.
41. Lanternari, 1963; Hall, 2013, p. 3; see also Barkun, 1986.
42. Hall, 2009.
43. For the Xhosa, see Peires, 1989, location 1106–1108; more generally, see Hall, 2009.
44. Hall, 2009; see also Scott, 1990, p. 101, who notes how "throughout Europe and in Southeast Asia . . . there are long traditions of the return of a just king or religious savior, despite great differences in cultural and religious lineages."
45. Acts 2:41 English Standard Version.
46. Lawlor & Oulton, 1928, 3.37.3.
47. MacMullen, 1984, p. 29; cited in Stone, 2016.
48. E.g., Abgrall, 1999; cited in Anthony, 1999.
49. Stark, 1996.
50. Stark, 1984.
51. E. Barker, 1984.
52. Iannaccone, 2006 (I have relied on this paper substantially for this section), drawing on E. Barker, 1984.
53. E. Barker, 1984.
54. E. Barker, 1984, p. 8
55. Stark & Bainbridge, 1980; David A. Snow & Phillips, 1980; Le Roy Ladurie, 2016, location 847–849; for a review, see Robbins, 1988; Stark & Bainbridge, 1980.
56. Anthony, 1999, p. 435.
57. Stark, 1996, p. 178; see also Power, 2017.
58. Stark, 1996, chap. 8.
59. Iannaccone, 2006, p. 7.
60. Murray, 1974, p. 307.
61. For an elaboration and a critique of this view, see Abercrombie et al., 1980 (from which I partly draw here).
62. Marx & Engels, 1970.

63. For general criticisms of this view, see Abercrombie et al., 1980; Delumeau, 1977; Le Bras, 1955; Stark, 1999; K. Thomas, 1971.

64. Murray, 1974, p. 299.

65. Murray, 1974, p. 304

66. Murray, 1974, p. 305

67. Murray, 1974, p. 318

68. Murray, 1974, pp. 305, 320. Humber was far from being the only preacher to deplore his flock's resistance to Christian doctrine; for more examples, see Pettegree, 2014, pp. 27, 128, 135, 137.

69. Le Roy Ladurie, 2016, location 666–673; see also Ekelund, Tollison, Anderson, Hébert, & Davidson, 1996.

70. Cohn, 1970; Dickson, 2007.

71. Cohn, 1970, p. 80.

72. Cohn, 1970, p. 80.

73. Delumeau, 1977, p. 225; see also MacMullen, 1999; K. Thomas, 1971.

74. Le Roy Ladurie, 2016, locations 947–952, 985–987.

75. Murray, 1974, p. 318. The point that official religious doctrine is never well accepted, and loses to more intuitively compelling traditional practices, is a truism in the anthropology of religion. For more examples related to medieval Christianity, see, e.g., Ginzburg, 2013.

76. Murray, 1974, pp. 307, 320.

77. On the case of Hinduism and the caste system, see Berreman, 1971; Harper, 1968; Juergensmeyer, 1980; Khare, 1984; Mencher, 1974.

78. Scott, 1990, 2008.

79. Scott, 2008, p. 29; for slaves in the antebellum South in particular, see Genovese, 1974.

CHAPTER 9

1. Hitler (1339); available at Project Gutenberg, http://gutenberg.net.au/ebooks02/0200601.txt. Quotes found in "Propaganda in Nazi Germany," *Wikipedia*, https://en.wikipedia.org/wiki/Propaganda_in_Nazi_Germany (both accessed November 23, 2018).

2. Kershaw, 1983a, p. 191.

3. Voigtländer & Voth, 2015.

4. For a different example of this process, see Bursztyn, Egorov, & Fiorin, 2019.

5. Adena, Enikolopov, Petrova, Santarosa, & Zhuravskaya, 2015, p. 1885.

6. Kershaw, 1983a, p. 191.

7. Kershaw, 1983a; Kuller, 2015.

8. Kershaw, 1983a, p. 188.

NOTES TO CHAPTER 9

9. Salter, 1983.
10. Stout, 2011, pp. 4, 31; see also Kallis, 2008; Kershaw, 1983a.
11. Kershaw, 1983a, 1987.
12. Kershaw, 1983a, p. 199.
13. See Mawson, 2012, p. 141.
14. Mawson, 2012, p. 141.
15. Stout, 2011.
16. Kershaw, 1983a, p. 200.
17. Brandenberger, 2012.
18. Davies, 1997, pp. 6–7. Russians also created a treasure trove of amazing jokes poking fun at the regime; see, for instance, "Russian political jokes," Wikipedia, https://en.wikipedia.org/wiki/Russian_political_jokes#Communism (accessed March 28, 2019).
19. Rose, Mishler, & Munro, 2011; B. Silver, 1987.
20. Peisakhin & Rozenas, 2018.
21. Wang, 1995.
22. Wang, 1995, p. 277.
23. X. Chen & Shi, 2001; see also Gang & Bandurski, 2011.
24. Osnos, 2014, location 2330–2333.
25. Osnos, 2014, location 3965–3966.
26. Huang, 2017.
27. Osnos, 2014, location 4657–4663.
28. Roberts, 2018, p. 218.
29. King, Pan, & Roberts, 2017.
30. Márquez, 2016, pp. 137–138, citing Aguilar, 2009; Pfaff, 2001; Tismaneanu, 1989.
31. For some examples of this, see Blakeslee, 2014; Petrova & Yanagizawa-Drott, 2016.
32. Kershaw, 1987, p. 80.
33. Demick, 2010.
34. Osnos, 2014, location 606–609.
35. Ezra Klein, "Trump has given North Korea 'the greatest gift ever,'" *Vox*, January 2, 2018, https://www.vox.com/2017/12/21/16803216/north-korea-trump-war.
36. Reviewed in J. J. Kennedy, 2009.
37. "China lifting 800 million people out of poverty is historic: World Bank," *Business Standard*, October 13, 2017, https://www.business-standard.com/article/international/china-lifting-800-million-people-out-of-poverty-is-historic-world-bank-117101300027_1.html; on the use of carrots by the Nazi, see the contested Aly, 2007.
38. Kershaw, 1983a, p. 196.

39. "Cost of Election," OpenSecrets.org, https://www.opensecrets.org/overview/cost.php (accessed July 6, 2018).

40. For references, see, e.g., O'Donnell & Jowett, 1992.

41. E.g., Lasswell, 1927.

42. Name given by Klapper, 1960, based on research such as Hovland, 1954; Lazarsfeld et al., 1948.

43. Klapper, 1960, p. 15, cited in Arceneaux & Johnson, 2013.

44. See, e.g., Iyengar & Kinder, 1987; Gamson, 1992.

45. Arceneaux & Johnson, 2013.

46. Lenz, 2009.

47. See Lenz, 2013; see also, e.g., Broockman & Butler, 2017; Carlsson, Dahl, & Rooth, 2015.

48. Berelson, Lazarsfeld, McPhee, & McPhee, 1954; Huckfeldt, Pietryka, & Reilly, 2014; Katz, 1957.

49. Chiang & Knight, 2011; Ladd & Lenz, 2009.

50. Kalla & Broockman, 2018, which also included some data from studies that used convincing identification strategies, approximating randomized experiments.

51. See also Broockman & Green, 2014; Durante & Gutierrez, 2014; S. J. Hill, Lo, Vavreck, & Zaller, 2013.

52. See references in Kalla & Broockman, 2018; see also Bekkouche & Cagé, 2018.

53. Kalla and Broockman did note a few exceptions to this pattern (although they might also be statistical flukes). When a campaign manages to provide a surprising piece of information to targeted voters for whom this information is particularly relevant, it might have a small effect. For example, in one study the campaign had managed to target pro-choice voters (people who oppose some restrictions on abortion) and sent mailers informing them that one of the candidates, whom they would have expected to be pro-choice, was in fact not (Rogers & Nickerson, 2013). This piece of information had a small but significant effect on these voters. What the campaign had done, essentially, was substituting itself for the media.

54. Carole Cadwalladr, "The great British Brexit robbery: How our democracy was hijacked," *Guardian*, May 7, 2017, https://www.theguardian.com/technology/2017/may/07/the-great-british-brexit-robbery-hijacked-democracy.

55. Matz, Kosinski, Nave, & Stillwell, 2017.

56. The quotes are from Evan Halper, "Was Cambridge Analytica a digital Svengali or snake-oil salesman?," *Los Angeles Times*, March 21, 2018, https://www.latimes.com/politics/la-na-pol-cambridge-analytica-20180321-story.html; the number estimates are from Mats Stafseng Einarsen (@matseinarsen), thread starting with "The Facebook + Cambridge Analytica thing is a trainwreck on multiple levels . . . ," Twitter, March 20, 2018, 9:44 a.m., https://twitter.com/matseinarsen/status/976137451025698821; see also Allan Smith, "There's an open secret about Cambridge

Analytica in the political world: It doesn't have the 'secret sauce' it claims," *Business Insider Australia*, March 22, 2018, https://www.businessinsider.com.au/cambridge-analytica-facebook-scandal-trump-cruz-operatives-2018-3; David Graham, "Not even Cambridge Analytica believed its hype," *Atlantic*, March 20, 2018, https://www.theatlantic.com/politics/archive/2018/03/cambridge-analyticas-self-own/556016/; Stephen Armstrong, "Cambridge Analytica's 'mindfuck tool' could be totally useless," *Wired*, March 22, 2018, https://www.wired.co.uk/article/cambridge-analytica-facebook-psychographics; Brian Resnick, "Cambridge Analytica's 'psychographic microtargeting': What's bullshit and what's legit," *Vox*, March 26, 2018, https://www.vox.com/science-and-health/2018/3/23/17152564/cambridge-analytica-psychographic-microtargeting-what.

57. Gelman, Goel, Rivers, & Rothschild, 2016.
58. Gelman & King, 1993, p. 409.
59. Barabas & Jerit, 2009, but it can create gaps in who knows what; Nadeau, Nevitte, Gidengil, & Blais, 2008; and see Roberts, 2018, for how this is used in China.
60. Ladd, 2011.
61. Besley, Burgess, & others, 2002; Snyder & Strömberg, 2010; Strömberg, 2004.
62. Peter Kafka and Rani Molla, "2017 was the year digital ad spending finally beat TV," *Vox*, December 4, 2017, https://www.recode.net/2017/12/4/16733460/2017-digital-ad-spend-advertising-beat-tv. This section is largely based on work reviewed by DellaVigna & Gentzkow, 2010; Tellis, 2003.
63. Lewis & Rao, 2013.
64. Aaker & Carman, 1982; see also Lodish et al., 1995.
65. Y. Hu, Lodish, & Krieger, 2007.
66. Ackerberg, 2001; Tellis, 1988; Tellis, Chandy, & Thaivanich, 2000.
67. Ackerberg, 2001, p. 318.
68. M. Thomas, in press.
69. Amos et al., 2008.
70. Lull & Bushman, 2015; Wirtz, Sparks, & Zimbres, 2018.
71. Nestlé, "Management Report 2005," https://www.nestle.com/asset-library/documents/library/documents/annual_reports/2005-management-report-en.pdf; Nestlé, "Management Report 2006," https://www.nestle.com/asset-library/documents/library/documents/annual_reports/2006-management-report-en.pdf; Nestlé, "Management Report 2007," https://www.nestle.com/asset-library/documents/library/documents/annual_reports/2007-management-report-en.pdf (all accessed May 25, 2019).
72. Christophe Cornu, "A new coffee for the USA from Nestlé Nespresso," Nestlé Investor Seminar 2014, available at https://www.slideshare.net/Nestle_IR/nespresso-35442357 (accessed May 25, 2019).

73. Nespresso, "Brand related," https://www.nestle-nespresso.com/about-us/faqs/brand-related (accessed May 25, 2019).

74. Van Doorn & Miloyan, 2017.

75. Tellis, 2003, p. 32; see also Blake, Nosko, & Tadelis, 2015. For some potential exceptions explained as solutions to coordination problems, see Chwe, 2001.

CHAPTER 10

1. Peter Schroeder, "Poll: 43 percent of Republicans believe Obama is a Muslim," The Hill, September 13, 2017, http://thehill.com/blogs/blog-briefing-room/news/253515-poll-43-percent-of-republicans-believe-obama-is-a-muslim.

2. Haifeng Huang, "In China, rumors are flying about David Dao's alleged $140 million settlement from United Airlines," Washington Post, May 10, 2017, https://www.washingtonpost.com/news/monkey-cage/wp/2017/05/10/in-china-rumors-are-flying-about-david-daos-140-million-settlement-from-united-airlines/.

3. Danny Cevallos, "United Airlines must have paid big bucks for Dr. Dao's silence," CNN, May 1, 2017, https://edition.cnn.com/2017/04/28/opinions/united-airlines-settlement-cevallos/index.html.

4. E. Morin, 1969.

5. Sinha, 1952.

6. Prasad, 1935.

7. Weinberg & Eich, 1978.

8. Allport & Postman, 1947.

9. Rosnow, 1991, p. 484.

10. Chorus, 1953, p. 314.

11. E.g., Naughton, 1996; P. A. Turner, 1992.

12. Pound & Zeckhauser, 1990.

13. What seems like a low hit rate (half of the rumors being false) is still very valuable because takeovers are rare. If we assume the odds of any given company, at any given time, being taken over are essentially zero, then hearing a rumor puts these odds at 50 percent, a massive change.

14. DiFonzo & Bordia, 2007.

15. DiFonzo, 2010.

16. DiFonzo, 2010, table 6.2, p. 146.

17. Caplow, 1947.

18. See also Knapp, 1944.

19. Caplow, 1947, p. 301.

20. Because our mechanisms of open vigilance are mostly in the business of accepting messages that would otherwise have been rejected, their inactivation doesn't raise significant risks of accepting harmful messages; see chapter 5.

21. For an application to the spread of bullshit, see Petrocelli, 2018.
22. Caplow, 1947.
23. Diggory, 1956.
24. Shibutani, 1966, p. 76.
25. Weinberg & Eich, 1978, p. 30.
26. Sperber, 1997.
27. Kay, 2011, p. 185.
28. Gwynn Guilford, "The dangerous economics of racial resentment during World War II," *Quartz*, February 13, 2018, https://qz.com/1201502/japanese-internment-camps-during-world-war-ii-are-a-lesson-in-the-scary-economics-of-racial-resentment/.
29. "Trump remains unpopular; Voters prefer Obama on SCOTUS pick," Public Policy Polling, December 9, 2016 https://www.publicpolicypolling.com/wp-content/uploads/2017/09/PPP_Release_National_120916.pdf.
30. Nation Pride comment on Google Review, https://www.google.com/search?q=comet+ping+pong&oq=comet+ping+pong&aqs=chrome..69i57j35i39j69i60j69i61j0l2.183j0j7&sourceid=chrome&ie=UTF-8#lrd=0x89b7c9b98f61ad27:0x81a8bf734dd1c58f,1,,, (accessed March 10, 2018).
31. Kanwisher, 2000.
32. Sperber, 1994.
33. NASA, public domain, https://commons.wikimedia.org/wiki/File:Cydonia_medianrp.jpg; and grendelkhan, https://www.flickr.com/photos/grendelkhan/119929591 (accessed June 18, 2019), CC BY-SA.
34. Wohlstetter, 1962, p. 368.
35. Boyer & Parren, 2015.
36. van Prooijen & Van Vugt, 2018.
37. Vosoughi, Roy, & Aral, 2018.
38. Boyer & Parren, 2015; Dessalles, 2007.
39. Donovan, 2004, p. 6.
40. Note that the same remarks apply to the spreading of "metarumors," rumors about the rumors—e.g., "Do you know that people are saying that Jewish shopkeepers are kidnapping young girls!" In this case as well, we should wonder what we would do if we really believed that people were intuitively believing this rumor, or what we would say to someone who is spreading it.

CHAPTER 11

1. Inspired by Buckner, 1965, p. 56.
2. Inspired by E. Morin, 1969, p. 113.
3. E. Morin, 1969, p. 113 (my translation); for an experimental replication of this phenomenon, see Altay, Claidière, & Mercier, submitted.

4. Aikhenvald, 2004, p. 43. Another source (the World Atlas of Language Structures sample) suggests 57 percent.

5. Aikhenvald, 2004, p. 43.

6. Aikhenvald, 2004, p. 26

7. Altay & Mercier, submitted; see also Vullioud et al., 2017.

8. On the importance of getting credit for communicated information, see Dessalles, 2007.

9. Four-year-olds seem able to draw this type of inference; see Einav & Robinson, 2011. On the whole, people seem to be quite sensitive to the attribution of proper sources, as they realize that this is important to grant proper credit or blame when the information turns out to be good or not; see I. Silver & Shaw, 2018.

10. Donovan, 2004, pp. 33ff.

11. "The royal family are bloodsucking alien lizards—David Icke," *Scotsman*, January 30, 2006, https://www.scotsman.com/news/uk-news/the-royal-family-are-bloodsucking-alien-lizards-david-icke-1-1103954.

12. Bod Drogin and Tom Hamburger, "Niger uranium rumors wouldn't die," *Los Angeles Times*, February 17, 2006, http://articles.latimes.com/2006/feb/17/nation/na-niger17.

13. Drogin and Hamburger, "Niger uranium rumors wouldn't die."

14. As a first approximation, in fact things are more complicated, as vetting of sources can add some independence: if many people agree a given source is reliable, it might be a good sign that it is indeed reliable (Estlund, 1994).

15. Dalai Lama (@DalaiLama), "Because of the great differences in our ways of thinking, it is inevitable that we have different religions and faiths. Each has its own beauty. And it is much better that we live together on the basis of mutual respect and mutual admiration," Twitter, February 26, 2018, 2:30 a.m., https://twitter.com/DalaiLama/status/968070699708379143?s=03).

16. I owe all of my information about the Duna to personal communications with her; see San Roque & Loughnane, 2012.

17. Rumsey & Niles, 2011.

18. See Boyer, 2001.

19. Baumard & Boyer, 2013b; Sperber, 1997.

20. Schieffelin, 1995.

21. In fact, the problem was present from the start. As the example of religious beliefs illustrates, even relatively simple cultures have chains that are difficult to trace.

22. Gloria Origgi, "Say goodbye to the information age: It's all about reputation now," *Aeon*, March 14, 2018, https://aeon.co/ideas/say-goodbye-to-the-information-age-its-all-about-reputation-now. See also Origgi, 2017.

23. E.g., Altay & Mercier, submitted; Mercier, Majima, Claidière, & Léone, submitted.

CHAPTER 12

1. Paul Wright "An innocent man speaks: PLN interviews Jeff Deskovic," *Prison Legal News*, August 15, 2013, https://www.prisonlegalnews.org/news/2013/aug/15/an-innocent-man-speaks-pln-interviews-jeff-deskovic/.
2. This section benefited from the review of Kassin & Gudjonsson, 2004.
3. Gudjonsson, Fridrik Sigurdsson, & Einarsson, 2004; Gudjonsson, Sigurdsson, Bragason, Einarsson, & Valdimarsdottir, 2004.
4. Gudjonsson & Sigurdsson, 1994; Sigurdsson & Gudjonsson, 1996.
5. "False confessions and recording of custodial interrogations," The Innocence Project, https://www.innocenceproject.org/causes/false-confessions-admissions/ (accessed April 4, 2018).
6. Kassin & Neumann, 1997.
7. Drizin & Leo, 2003.
8. Quoted in Kassin & Gudjonsson, 2004, p. 36.
9. Gudjonsson, Sigurdsson, et al., 2004.
10. Radelet, Bedau, & Putnam, 1994.
11. Kassin & Gudjonsson, 2004, p. 50.
12. Jonathan Bandler, "Deskovic was 'in fear for my life' when he confessed," *Lohud*, October 21, 2014, https://www.lohud.com/story/news/local/2014/10/21/jeffrey-deskovic-wrongful-conviction-putnam-county-daniel-stephens/17680707/.
13. See Kassin & Gudjonsson, 2004.
14. Kassin, Meissner, & Norwick, 2005. Interrogators are also routinely taught to rely on cues known to be completely useless, such as gaze aversion; see chapter 9.
15. Kassin & Wrightsman, 1980.
16. Indeed, in adversarial systems, the prosecution is under no obligation to stress the existence of such pressures.
17. Kassin & Wrightsman, 1980.
18. Futaba & McCormack, 1984.
19. Parker & Jaudel, 1989. The situation is arguably worse in China; see, e.g., "'My hair turned white': Report lifts lid on China's forced confessions," *Guardian*, April 12, 2018, https://www.theguardian.com/world/2018/apr/12/china-forced-confessions-report.
20. Gross, 2018, p. 21.
21. Gross, 2018, p. 22.
22. Evans-Pritchard, 1937, pp. 22–23.
23. Alternatively, Aleksander might feel shame if enough people think he is guilty, whether he is guilty or not, and this shame might make a confession more tempting, even rational, as what matters for shame isn't really what we have done, but what people

believe we have done (T. E. Robertson, Sznycer, Delton, Tooby, & Cosmides, 2018; Sznycer, Schniter, Tooby, & Cosmides, 2015; Sznycer et al., 2018). Whether or not we are guilty, the confession should help us redeem ourselves in the eyes of the people who believe we are guilty.

24. It is also possible (even more likely) that beliefs in supernatural evil acts are initially attributed to imaginary agents (such as ancestors), which is easier to suggest than (live) human agents, and then these acts become more conceivable as something real agents could confess to.

25. Hutton, 2017, p. 59.
26. Ward, 1956.
27. Ardener, 1970.
28. Burridge, 1972.
29. Hutton, 2017, p. 37.
30. Willis, 1970, p. 130; see also R. Brain, 1970.
31. T. E. Robertson et al., 2018; Sznycer et al., 2015.
32. Lévi-Strauss, 1967.
33. Macfarlane, 1970, p. 91.
34. Morton-Williams, 1956, p. 322.
35. Evans-Pritchard, 1937, p. 48.
36. Miguel, 2005.
37. Julian Ryall, "The incredible Kim Jong-il and his amazing achievements," *Telegraph*, January 31 2011, https://www.telegraph.co.uk/news/worldnews/asia/northkorea/8292848/The-Incredible-Kim-Jong-il-and-his-Amazing-Achievements.html.
38. All quotes cited in Hassig & Oh, 2009, p. 57.
39. In addition to the references above, see AFP, "N. Korea leader sets world fashion trend: Pyongyang," *France 24*, April 7, 2010, https://web.archive.org/web/20111219011527/http://www.france24.com/en/20100407-nkorea-leader-sets-world-fashion-trend-pyongyang.
40. Wedeen, 2015; Sebestyen, 2009; Harding, 1993; Karsh & Rautsi, 2007. This list is largely owed to Márquez, 2018; and Svolik, 2012, p. 80. Kim Jong-un, Kim Jong-il's son and successor, seems intent on topping his father in terms of absurd praise; see Fifield, 2019.
41. Hassig & Oh, 2009.
42. Márquez, 2018.
43. Leese, 2011, p. 168, found in Márquez, 2018.
44. Technically, these groups are referred to as coalitions; see Tooby, Cosmides, & Price, 2006.
45. E.g., Delton & Cimino, 2010.
46. Personal communication, July 4, 2016; see also Kurzban & Christner, 2011.

47. Jerry Coyne, "The University of Edinburgh and the John Templeton Foundation royally screw up evolution and science (and tell arrant lies) in an online course," Why Evolution is True, https://whyevolutionistrue.wordpress.com/2018/03/25/the-university-of-edinburgh-and-the-john-templeton-foundation-royally-screw-up-evolution-and-science-and-tell-arrant-lies-in-an-online-course/ (accessed April 12, 2018).

48. Jerry Coyne, "A postmodern holiday: Recent nonsense from the humanities," Why Evolution is True, https://whyevolutionistrue.wordpress.com/2017/01/10/a-postmodern-holiday-recent-nonsense-from-the-humanities/ (accessed April 12, 2018). For examples on the other side of the intellectual spectrum, see Alice Dreger, "Why I escaped the 'Intellectual Dark Web,'" *Chronicle of Higher Education*, May 11, 2018, https://www.chronicle.com/article/Why-I-Escaped-the/243399.

49. Tibor Machan, "Tax slavery," Mises Institute, March 13, 2000, https://mises.org/library/tax-slavery; and Rothbard, 2003, p. 100.

50. Fresco, 1980.

51. Cahal Milmo, "Isis video: 'New Jihadi John' suspect Siddhartha Dhar is a 'former bouncy castle salesman from east London,'" *Independent*, January 4, 2016, https://www.independent.co.uk/news/uk/home-news/isis-video-new-jihadi-john-suspect-is-a-former-bouncy-castle-salesman-from-east-london-a6796591.html (found in Roy, 2016).

52. Márquez, 2018, inspired by Winterling, 2011.

53. On the incremental nature of burning bridges beliefs, see, e.g., Josiah Hesse, "Flat Earthers keep the faith at Denver conference," *Guardian*, November 18, 2018, https://www.theguardian.com/us-news/2018/nov/18/flat-earthers-keep-the-faith-at-denver-conference (on flat-earthers) or Ben Sixsmith, "The curious case of Ron Unz," *Spectator USA*, September 15, 2018, https://spectator.us/ron-unz/ (on negationists).

54. Gudjonsson, 2003.

CHAPTER 13

1. A lively history of the humoral theory can be found in Arika, 2007.

2. Wootton, 2006, p. 37.

3. P. Brain, 1986, pp. 26–27, 33.

4. For an early example of blood libel, see Hugo Mercier, "Blatant bias and blood libel," *International Cognition and Culture Institute*, January 28, 2019, http://cognitionandculture.net/blog/hugo-merciers-blog/blatant-bias-and-blood-libel/.

5. See, e.g., Horowitz, 2001, pp. 75ff.

6. Alison Flood, "Fake news is 'very real' word of the year for 2017," *Guardian*, November 2, 2017, https://www.theguardian.com/books/2017/nov/02/fake-news-is-very-real-word-of-the-year-for-2017.

7. Andrew Grice, "Fake news handed Brexiteers the referendum—and now they have no idea what they're doing," *Independent*, January 18, 2017, https://www.independent.co.uk/voices/michael-gove-boris-johnson-brexit-eurosceptic-press-theresa-may-a7533806.html; Aaron Blake, "A new study suggests fake news might have won Donald Trump the 2016 election," *Washington Post*, April 3, 2018 https://www.washingtonpost.com/news/the-fix/wp/2018/04/03/a-new-study-suggests-fake-news-might-have-won-donald-trump-the-2016-election/.

8. Larson, 2018.

9. See, e.g., John Lichfield, "Boris Johnson's £350m claim is devious and bogus. Here's why," *Guardian*, September 18, 2017, https://www.theguardian.com/commentisfree/2017/sep/18/boris-johnson-350-million-claim-bogus-foreign-secretary.

10. See, e.g., Robert Darnton, "The true history of fake news," *New York Review of Books*, February 13, 2017, http://www.nybooks.com/daily/2017/02/13/the-true-history-of-fake-news/.

11. Craig Silverman, "This analysis shows how viral fake election news stories outperformed real news on Facebook," BuzzFeed, November 16, 2016, https://www.buzzfeed.com/craigsilverman/viral-fake-election-news-outperformed-real-news-on-facebook.

12. E.g., Del Vicario, Scala, Caldarelli, Stanley, & Quattrociocchi, 2017; Zollo et al., 2017.

13. The citation is from a translation of Normal Lewis Torrey (1961, p. 278). The French is "Certainement qui est en droit de vous rendre absurde est en droit de vous rendre injuste." See Voltaire, *Œuvres complètes*, available at https://fr.wikisource.org/wiki/Page:Voltaire_-_%C5%92uvres_compl%C3%A8tes_Garnier_tome25.djvu/422 (accessed May 22, 2019).

14. Wootton, 2006.

15. The anthropological references were located thanks to the Human Relations Area Files (HRAF). See Epler, 1980; Miton, Claidière, & Mercier, 2015; Murdock, Wilson, & Frederick, 1978.

16. Miton et al., 2015.

17. Horowitz, 2001.

18. Zipperstein, 2018, p. 89.

19. Zipperstein, 2018, p. 94.

20. Shibutani, 1966, p. 113, approving of; R. H. Turner, 1964; see also Horowitz, 2001, p. 86.

21. Guess, Nyhan, & Reifler, 2018.

22. Fourney, Racz, Ranade, Mobius, & Horvitz, 2017.

23. Benedict Carey, "'Fake news': Wide reach but little impact, study suggests," *New York Times*, January 2, 2018, https://www.nytimes.com/2018/01/02/health/fake-news-conservative-liberal.html, and Guess et al., 2018.

24. See, e.g., Druckman, Levendusky, & McLain, 2018. On the fact that internet use does not explain Republican votes, see also Boxell, Gentzkow, & Shapiro, 2018.

25. Nyhan et al., 2017. For a similar effect in another domain, see Hopkins et al., 2019.

26. J. W. Kim & Kim, in press; see also Benkler, Faris, & Roberts, 2018.

27. Malle, Knobe, & Nelson, 2007, study 3.

28. Lloyd & Sivin, 2002.

29. P. Brain, 1986.

30. See also Vargo, Guo, & Amazeen, 2018.

31. Craig Silverman, "Here are 50 of the biggest fake news hits on Facebook from 2016," BuzzFeed, December 16, 2016, https://www.buzzfeed.com/craigsilverman/top-fake-news-of-2016, data available at: https://docs.google.com/spreadsheets/d/1sTkRkHLvZp9XlJOynYMXGslKY9fuB_e-2mrxqgLwvZY/edit#gid=652144590 (accessed April 24, 2018).

32. Craig Silverman, Jane Lytvynenko, & Scott Pham, "These are 50 of the biggest fake news hits on facebook in 2017," BuzzFeed, December 28, 2017, https://www.buzzfeed.com/craigsilverman/these-are-50-of-the-biggest-fake-news-hits-on-facebook-in.

33. Allcott & Gentzkow, 2017; Guess, Nagler, & Tucker, 2019.

34. Grinberg, Joseph, Friedland, Swire-Thompson, & Lazer, 2019; Guess et al., 2019.

35. See the Buzzfeed articles referenced in notes 31 and 32.

36. Acerbi, 2019. On the lack of partisanship effects, see Pennycook & Rand, 2018. Another potential explanation for the sharing of fake news is a "need for chaos": it seems some people share fake news left and right, reflecting a more general contestation of the existing system (Petersen, Osmundsen, & Arceneaux, 2018).

37. Sadler & Tesser, 1973.

38. For a review, see Tesser, 1978.

39. Myers & Bach, 1974.

40. Isenberg, 1986; Vinokur, 1971.

41. See earlier in the chapter on echo chambers, as well as, for journalistic references, Mostafa El-Bermawy, "Your filter bubble is destroying democracy," Wired, November 18, 2016, https://www.wired.com/2016/11/filter-bubble-destroying-democracy/; Christopher Hooton, "Social media echo chambers gifted Donald Trump the presidency," Independent, November 10, 2016, https://www.independent.co.uk/voices/donald-trump-president-social-media-echo-chamber-hypernormalisation-adam-curtis-protests-blame-a7409481.html.

42. Sunstein, 2018.

43. E.g., Jonathan Haidt, & Sam Abrams, "The top 10 reasons American politics are so broken," Washington Post, January 7, 2015, https://www.washingtonpost.com/news/wonk/wp/2015/01/07/the-top-10-reasons-american-politics-are-worse-than-ever.

44. El-Bermawy, "Your filter bubble is destroying democracy."

45. Fiorina, Abrams, & Pope, 2005; see also Desmet & Wacziarg, 2018; Jeffrey Jones, "Americans' identification as independents back up in 2017," Gallup, January 8, 2018, http://news.gallup.com/poll/225056/americans-identification-independents-back-2017.aspx.

46. See the data from the GSS Data Explorer available at https://gssdataexplorer.norc.org/trends/Politics?measure=polviews_r (accessed April 25, 2018).

47. "Political polarization in the American public," Pew Research Center, June 12, 2014, http://www.people-press.org/2014/06/12/political-polarization-in-the-american-public/.

48. "Political polarization in the American public."

49. Shore, Baek, & Dellarocas, 2018.

50. "Political polarization in the American public."

51. It might also partly be the outcome of a sampling bias; see Cavari & Freedman, 2018.

52. Jason Jordan, "Americans are getting smarter about politics in at least one important way," *Washington Post*, February 7, 2018, https://www.washingtonpost.com/news/monkey-cage/wp/2018/02/07/americans-are-getting-smarter-about-politics-in-at-least-one-important-way/?utm_term=.89ff43081c86.

53. See Iyengar, Lelkes, Levendusky, Malhotra, & Westwood, 2019. On the relationship between sorting and affective polarization, see Webster & Abramowitz, 2017. Still, even affective polarization might not be as problematic as some fear; see Klar, Krupnikov, & Ryan, 2018; Tappin & McKay, 2019; Westwood, Peterson, & Lelkes, 2018.

54. Elizabeth Dubois & Grant Blank, "The myth of the echo chamber," The Conversation, March 8, 2018, https://theconversation.com/the-myth-of-the-echo-chamber-92544.

55. Gentzkow & Shapiro, 2011.

56. Fletcher & Nielsen, 2017.

57. Guess, 2016; see also Flaxman, Goel, & Rao, 2016; R. E. Robertson et al., 2018.

58. Dubois & Blank, "The myth of the echo chamber"; see also Puschmann, 2018.

59. Allcott, Braghieri, Eichmeyer, & Gentzkow, 2019.

60. Beam, Hutchens, & Hmielowski, 2018; see also Jo, 2017.

61. Boxell, Gentzkow, & Shapiro, 2017; see also Andrew Guess, Benjamin Lyons, Brendan Nyhan, & Jason Reifler, "Why selective exposure to like-minded congenial political news is less prevalent than you think," *Medium*, February 13, 2018, https://medium.com/trust-media-and-democracy/avoiding-the-echo-chamber-about-echo-chambers-6e1f1a1a0f39 (accessed April 26, 2018).

62. See, e.g., Crowell & Kuhn, 2014.

63. Zipperstein, 2018, p. 29.

64. Darnton, "The True History of Fake News."
65. Kaplan, 1982.

CHAPTER 14

1. See, e.g., Shtulman, 2017.
2. Denis Dutton, "The Bad Writing Contest," denisdutton.com, http://www.denisdutton.com/bad_writing.htm (accessed June 8, 2018).
3. This is an honest attempt at translating the following: "Pour couper court, je dirai que la nature se spécifie de n'être pas une, d'où le procédé logique pour l'aborder. Par le procédé d'appeler nature ce que vous excluez du fait même de porter intérêt à quelque chose, ce quelque chose se distinguant d'être nommé, la nature ne se risque à rien qu'à s'affirmer d'être un pot-pourri de hors-nature" (Lacan, 2005, p. 12).
4. "When Americans say they believe in God, what do they mean?," Pew Research Center, April 25, 2018, http://www.pewforum.org/2018/04/25/when-americans-say-they-believe-in-god-what-do-they-mean/.
5. E.g., "Mixed messages about public trust in science," Pew Research Center, December 8, 2017, http://www.pewinternet.org/2017/12/08/mixed-messages-about-public-trust-in-science/.
6. Cited in McIntyre, 2018, p. 142.
7. Sperber, 1997.
8. Boyer, 2001; Sperber, 1975.
9. Boyer, 2001.
10. Boyer, 2001.
11. Greene, 1990.
12. Barrett, 1999.
13. Barrett & Keil, 1996.
14. Barrett, 1999, p. 327.
15. Barrett & Keil, 1996.
16. See also Barlev, Mermelstein, & German, 2017, 2018.
17. McCloskey, Washburn, & Felch, 1983.
18. See Dennett, 1995.
19. My translation from Lévi-Strauss, 1986, cited in "Jacques Lacan," Wikipedia, https://fr.wikipedia.org/wiki/Jacques_Lacan (accessed May 15. 2018).
20. "Jacques Lacan," Wikipedia.
21. On plausibility, see Collins et al., 2018.
22. Boyer & Parren, 2015.
23. Thomas Mackie, "Lethal pig virus similar to SARS could strike humans," InfoWars, May 15, 2018, https://www.infowars.com/lethal-pig-virus-similar-to-sars-could-strike-humans/; "Experts: MH370 pilot was on murder suicide mission,"

InfoWars, May 15, 2018, https://www.infowars.com/experts-mh370-pilot-was-on-murder-suicide-mission/.

24. Paul Joseph Watson, "Finland: 93% of migrant sex crimes committed by migrants from Islamic countries," InfoWars, May 15, 2018, https://www.infowars.com/finland-93-of-migrant-sex-crimes-committed-by-migrants-from-islamic-countries/; "Watch live: Turkey announces launch of worldwide Jihad, withdraws ambassadors from US/Israel," InfoWars, May 15, 2018, https://www.infowars.com/watch-live-soros-shuts-down-offices-in-repressive-hungary/; "Video exposes the suicide of Europe," InfoWars, May 15, 2018, https://www.infowars.com/video-exposes-the-suicide-of-europe/.

25. CNBC, "The George Soros foundation says it is being forced to close its offices in Hungary," InfoWars, May 15, 2018, https://www.infowars.com/the-george-soros-foundation-says-it-is-being-forced-to-close-its-offices-in-hungary/. The video was available there until YouTube revoked the InfoWars account: https://www.youtube.com/watch?v=t41lx_ur4Y8 (accessed May 16, 2018).

26. The products can be found in the InfoWars store: https://www.infowarsstore.com/survival-shield-x-2-nascent-iodine.html, https://www.infowarsstore.com/preparedness/emergency-survival-foods.html, https://www.infowarsstore.com/preparedness/nuclear-and-biological/radiological-rad-replacement-filter.html (accessed May 16, 2018).

27. P. Brain, 1986, p. 33.

28. P. Brain, 1986, p. 90; Miton et al., 2015.

29. P. Brain, 1986, pp. 85, 89.

30. Cheatham, 2008.

31. Baumard & Boyer, 2013a; Baumard & Chevallier, 2015; Baumard, Hyafil, Morris, & Boyer, 2015.

32. Boyer & Baumard, 2018; see also Baumard et al., 2015.

33. Kenneth Doyle, "What happened to the people who died before Jesus was born?," Crux, August 24, 2015, https://cruxnow.com/church/2015/08/24/what-happened-to-the-people-who-died-before-jesus-was-born/.

34. R. Wright, 2009, who notes: "After Adam and Eve ate the forbidden fruit, according to Genesis, 'they heard the sound of the Lord God walking in the garden at the time of the evening breeze, and the man and his wife hid themselves from the presence of the Lord God among the trees of the garden.' Hiding may sound like a naive strategy to deploy against the omniscient God we know today, but apparently he wasn't omniscient back then. For 'the Lord God called to the man, and said to him, "Where are you?"'" (p. 103).

35. Eriksson, 2012, p. 748.

36. Weisberg, Keil, Goodstein, Rawson, & Gray, 2008.

37. On the importance of prestige, see Clauset, Arbesman, & Larremore, 2015; Goues et al., 2017; A. C. Morgan, Economou, Way, & Clauset, 2018.

38. L. T. Benjamin & Simpson, 2009; Griggs & Whitehead, 2015.
39. Arendt, 1963; Brown, 1965.
40. Perry, 2013, pp. 304ff. For a meta-analysis of Milgram's obedience experiments, see Haslam, Loughnan, & Perry, 2014.
41. Even more people claimed they were not really hurting the learner; see Hollander & Turowetz, 2017; and for new data on that point, see Perry, Brannigan, Wanner, & Stam, in press.
42. Milgram, 1974, p. 172.
43. Reicher, Haslam, & Smith, 2012.
44. Burger, Girgis, & Manning, 2011.
45. Perry, 2013, p. 310.
46. Blancke, Boudry, & Pigliucci, 2017.
47. Sokal & Bricmont, 1998.
48. Sperber & Wilson, 1995.
49. Honda, "Airbag inflator recall," https://www.honda.co.uk/cars/owners/airbag-recall.html.
50. Sperber, 2010.
51. Lacan, 1980.
52. Lacan, 1939.
53. Sokal & Bricmont, 1998, p. 34.
54. My translation from Lévi-Strauss, 1986, cited in "Jacques Lacan," *Wikipedia*.
55. Lacan, 1970, p. 193.
56. Milner, 1995; cited and translated in Sokal & Bricmont, 1998.
57. Goldman, 2001.
58. Peterson, 2002, p. 286.
59. Deepak Chopra (@DeepakChopra), "Mechanics of Manifestation: Intention, detachment, centered in being allowing juxtaposition of possibilities to unfold #CosmicConsciousness," Twitter, May 28, 2014, 2:24 a.m., https://twitter.com/deepakchopra/status/471582895622991872; Deepak Chopra (@DeepakChopra), "As beings of light we are local and nonlocal, time bound and timeless actuality and possibility #CosmicConsciousness," Twitter, May 5, 2014, 5:20 a.m., https://twitter.com/deepakchopra/status/463292121794224128. Fittingly, these tweets are drawn from a study on receptivity to bullshit: Pennycook, Cheyne, Barr, Koehler, & Fugelsang, 2015.

CHAPTER 15

1. Baumard et al., 2013.
2. Martin & Yurukoglu, 2017.
3. Ahler & Sood, 2018; Levendusky & Malhotra, 2015; Westfall, Van Boven, Chambers, & Judd, 2015.

NOTES TO CHAPTER 15

4. Yang et al., 2016.
5. Enders & Armaly, 2018.
6. Stroud & Lee, 2013.
7. Aaron Sharockman, "Fact-checking Fox, MSNBC, and CNN: PunditFact's network scorecards," Punditfact, September 16, 2014, http://www.politifact.com/punditfact/article/2014/sep/16/fact-checking-fox-msnbc-and-cnn-punditfacts-networ/.
8. DellaVigna & Kaplan, 2007; see also Martin & Yurukoglu, 2017.
9. Schroeder & Stone, 2015.
10. See also Hopkins & Ladd, 2014, who conclude that "the potential voters who tend to agree with Fox News's overall slant are those more likely to be influenced by the channel" (p. 129).
11. Gelman & King, 1993, p. 409.
12. Martin & Yurukoglu, 2017.
13. Moon, Krems, & Cohen, 2018.
14. McCullough, Swartwout, Shaver, Carter, & Sosis, 2016.
15. Foddy, Platow, & Yamagishi, 2009; Platow, Foddy, Yamagishi, Lim, & Chow, 2012.
16. Marina Lopes, "One way out: Pastors in Brazil converting gang members on YouTube," *Washington Post*, May 17, 2019, https://www.washingtonpost.com/world/the_americas/one-way-out-pastors-in-brazil-converting-gang-members-on-youtube/2019/05/17/be560746-614c-11e9-bf24-db4b9fb62aa2_story.html.
17. See Maurer, 1999.
18. Maurer, 1999, p. 4.
19. Braucher & Orbach, 2015.
20. From a contemporary newspaper report, cited by Braucher & Orbach, 2015, p. 256.
21. Mitnick & Simon, 2002, p. 26.
22. Braucher & Orbach, 2015, p. 263.
23. Braucher & Orbach, 2015, p. 249.
24. A very old ploy indeed: Pierre Ropert, "Histoires d'arnaques : Du mail du prince nigérian aux 'lettres de Jérusalem,'" France Culture, June 21, 2018, https://www.franceculture.fr/histoire/avant-les-mails-de-princes-nigerians-au-xviiieme-siecle-larnaque-aux-lettres-de-jerusalem.
25. E.g., "Crackdown on £8.4m African sting," *Scotsman*, March 2, 2003, https://www.scotsman.com/news/uk/crackdown-on-163-8-4m-african-sting-1-1382507 (accessed May 31, 2018).
26. Herley, 2012.
27. Berg, Dickhaut, & McCabe, 1995.
28. Charness & Dufwenberg, 2006; see also Schniter, Sheremeta, & Sznycer, 2013. For reviews on the efficacy of communication in economic games and social dilemmas, see Balliet, 2010; Sally, 1995.

29. Ostrom, Walker, & Gardner, 1992; see also Mercier, submitted.
30. E.g., Yamagishi, 2001.
31. Yamagishi, 2001.
32. Y. Chen, YeckehZaare, & Zhang, 2018.
33. Fershtman & Gneezy, 2001.
34. Burns, 2012.
35. Gupta, Mahmud, Maitra, Mitra, & Neelim, 2013; but see Glaeser, Laibson, Scheinkman, & Soutter, 2000.

CHAPTER 16

1. Brennan, 2012, p. 8.
2. G. Kim & Kwak, 2011.
3. Stenberg, 2013.
4. Castelain, Bernard, & Mercier, 2018.
5. Sodian, Thoermer, & Dietrich, 2006; Terrier, Bernard, Mercier, & Clément, 2016; VanderBorght & Jaswal, 2009.
6. J. Hu et al., 2015; T.J.H. Morgan, Laland, & Harris, 2015.
7. Kershaw, 1983a, p. 200.
8. See Barrows, 1981.
9. Stanley, 2015, p. 27.
10. Brennan, 2016.
11. Rousseau, 2002.
12. Veyne, 2002.
13. Slava Malamud (@SlavaMalamud), thread starting with "1/ So, here is what's happening in the KHL, for those who still can't quite grasp the banality of evil, Russian style . . ." Twitter, March 7, 2018, 7:57 p.m., https://twitter.com/slavamalamud/status/971595788315918336?lang=en.
14. For a model of this type of propaganda, see Márquez, 2018.
15. Dikötter, 2010.
16. Dikötter, 2010, locations 996–997.
17. E.g., Jeremy Diamond, "Trump sides with Putin over US intelligence," CNN, July 16, 2018, https://edition.cnn.com/2018/07/16/politics/donald-trump-putin-helsinki-summit/index.html.
18. Art Swift, "Putin's image rises in U.S., mostly among Republicans," Gallup, February 21, 2017, https://news.gallup.com/poll/204191/putin-image-rises-mostly-among-republicans.aspx (this poll predates the Helsinki summit, and so only reflects Trump's prior actions). More generally, see Lenz, 2013.
19. Stimson, 2004.
20. P. Benjamin & Shapiro, 1992; Stimson, 2004.

21. "Shifting public views on legal immigration into the U.S.," Pew Research Center, June 28, 2018, http://www.people-press.org/2018/06/28/shifting-public-views-on-legal-immigration-into-the-u-s/.

22. Wlezien, 1995; see also Stimson, 2004.

23. E.g., Horne, Powell, Hummel, & Holyoak, 2015; Nyhan & Reifler, 2015.

24. See, e.g., Goldacre, 2014.

25. Faasse, Chatman, & Martin, 2016; Fadda, Allam, & Schulz, 2015.

26. Chanel, et al., 2011.

27. E.g., Charlotte Gao, "HNA Group chairman's sudden death stokes conspiracy theories," *Diplomat*, July 5, 2018, https://thediplomat.com/2018/07/hna-group-chairmans-sudden-death-stokes-conspiracy-theories/; Rachel Lu, "Chinese conspiracy theorists of the world, unite!," *Foreign Policy*, May 11, 2015, https://foreignpolicy.com/2015/05/11/chinese-conspiracy-theorists-of-the-world-unite-hong-kong-banned-books/.

28. On how to build trust in society, see Algan, Cahuc, & Zilberberg, 2012.

29. On Pakistan, see "What is the wildest conspiracy theory pertaining to Pakistan?," *Herald*, June 19, 2015, https://herald.dawn.com/news/1153068.

REFERENCES

Aaker, D. A., & Carman, J. M. (1982). "Are you over-advertising?" *Journal of Advertising Research, 22*(4), 57–70.

Aarnio, K., & Lindeman, M. (2005). "Paranormal beliefs, education, and thinking styles." *Personality and Individual Differences, 39*(7), 1227–1236.

Abercrombie, N., Hill, S., & Turner, B. S. (1980). *The dominant ideology thesis.* London: Allen & Unwin.

Abgrall, J.-M. (1999). *Soul snatchers: The mechanics of cults.* New York: Algora.

Acerbi, A. (2019). "Cognitive attraction and online misinformation." *Palgrave Communications, 5*(1), 15.

Ackerberg, D. A. (2001). "Empirically distinguishing informative and prestige effects of advertising." *RAND Journal of Economics, 32*(2) 316–333.

Adena, M., Enikolopov, R., Petrova, M., Santarosa, V., & Zhuravskaya, E. (2015). "Radio and the rise of the Nazis in prewar Germany." *Quarterly Journal of Economics, 130*(4), 1885–1939.

Aguilar, P. (2009). "Whatever happened to Francoist socialization? Spaniards' values and patterns of cultural consumption in the post-dictatorial period." *Democratization, 16*(3), 455–484.

Ahler, D. J., & Sood, G. (2018). "The parties in our heads: Misperceptions about party composition and their consequences." *Journal of Politics, 80*(3), 964–981.

Aikhenvald, A. Y. (2004). *Evidentiality.* Oxford: Oxford University Press.

Aird, M. J., Ecker, U. K., Swire, B., Berinsky, A. J., & Lewandowsky, S. (2018). "Does truth matter to voters? The effects of correcting political misinformation in an Australian sample." *Royal Society Open Science, 5*(12), 180593.

Alexander, M., & Bruning, J. (2008). *How to break a terrorist: The U.S. interrogators who used brains, not brutality, to take down the deadliest man in Iraq.* New York: Free Press.

Algan, Y., Cahuc, P., & Zilberberg, A. (2012). *La Fabrique de la défiance : ... Et comment s'en sortir*. Paris: Albin Michel.

Allcott, H., Braghieri, L., Eichmeyer, S., & Gentzkow, M. (2019). *The welfare effects of social media*. NBER Working Paper No. 25514. Retrieved from https://www.nber.org/papers/w25514

Allcott, H., & Gentzkow, M. (2017). "Social media and fake news in the 2016 election." *Journal of Economic Perspectives*, 31(2), 211–236.

Allen, V. L. (1965). "Situational factors in conformity." *Advances in Experimental Social Psychology*, 2, 133–175.

Allport, G. W., & Postman, L. (1947). *The psychology of rumor*. Oxford: Henry Holt.

Altay, S., Claidière, N., & Mercier, H. (submitted). *Chain shortening in rumor transmission*.

Altay, S., & Mercier, H. (submitted). *I found the solution! How we use sources to appear competent*.

Aly, G. (2007). *Hitler's beneficiaries: Plunder, racial war, and the Nazi welfare state*. London: Macmillan.

Amos, C., Holmes, G., & Strutton, D. (2008). "Exploring the relationship between celebrity endorser effects and advertising effectiveness: A quantitative synthesis of effect size." *International Journal of Advertising*, 27(2), 209–234.

Analytis, P. P., Barkoczi, D., & Herzog, S. M. (2018). "Social learning strategies for matters of taste." *Nature Human Behaviour*, 2(6), 415–424.

Anderson, C., Brion, S., Moore, D. A., & Kennedy, J. A. (2012). "A status-enhancement account of overconfidence." *Journal of Personality and Social Psychology*, 103(4), 718–735.

Anthony, D. (1999). "Pseudoscience and minority religions: An evaluation of the brainwashing theories of Jean-Marie Abgrall." *Social Justice Research*, 12(4), 421–456.

Arceneaux, K., & Johnson, M. (2013). *Changing minds or changing channels? Partisan news in an age of choice*. Chicago: University of Chicago Press.

Ardener, E. (1970). "Witchcraft, economics, and the continuity of belief." In M. Douglas (Ed.), *Witchcraft confessions and accusations* (pp. 141–160). London: Routledge.

Arendt, H. (1963). *Eichmann in Jerusalem: A report on the banality of evil*. New York: Viking.

Arika, N. (2007). *Passions and tempers: A history of the humors*. New York: Harper Perennial.

Artés, J. (2013). "Do Spanish politicians keep their promises?" *Party Politics*, 19(1), 143–158.

Asch, S. E. (1956). "Studies of independence and conformity: A minority of one against a unanimous majority." *Psychological Monographs*, 70(9), 1–70.

Aveni, A. F. (1977). "The not-so-lonely crowd: Friendship groups in collective behavior." *Sociometry, 40*(1), 96–99.
Bago, B., & De Neys, W. (2019). "The smart System 1: Evidence for the intuitive nature of correct responding in the bat-and-ball problem." *Thinking & Reasoning, 25*(3), 257–299.
Bahrami, B., Olsen, K., Latham, P. E., Roepstorff, A., Rees, G., & Frith, C. D. (2010). "Optimally interacting minds." *Science, 329*(5995), 1081–1085.
Balliet, D. (2010). "Communication and cooperation in social dilemmas: A meta-analytic review." *Journal of Conflict Resolution, 54*(1), 39–57.
Barabas, J., & Jerit, J. (2009). "Estimating the causal effects of media coverage on policy-specific knowledge." *American Journal of Political Science, 53*(1), 73–89.
Barker, E. (1984). *The making of a Moonie: Choice or brainwashing?* Oxford: Blackwell.
Barker, J. (2014). *1381: The year of the Peasants' Revolt.* Cambridge, MA: Harvard University Press.
Barkow, J. H., Cosmides, L., & Tooby, J. (1992). *The adapted mind.* Oxford: Oxford University Press.
Barkun, M. (1986). *Disaster and the millennium.* Syracuse, NY: Syracuse University Press.
Barlev, M., Mermelstein, S., & German, T. C. (2017). "Core intuitions about persons coexist and interfere with acquired Christian beliefs about God." *Cognitive Science, 41*(53), 425–454.
Barlev, M., Mermelstein, S., & German, T. C. (2018). "Representational coexistence in the God concept: Core knowledge intuitions of God as a person are not revised by Christian theology despite lifelong experience." *Psychonomic Bulletin and Review, 25*(6) 1–9.
Barrett, J. L. (1999). "Theological correctness: Cognitive constraint and the study of religion." *Method and Theory in the Study of Religion, 11*(4), 325–339.
Barrett, J. L., & Keil, F. C. (1996). "Conceptualizing a nonnatural entity: Anthropomorphism in God concepts." *Cognitive Psychology, 31*(3), 219–247.
Barrows, S. (1981). *Distorting mirrors: Visions of the crowd in late nineteenth-century France.* New Haven, CT: Yale University Press.
Baumard, N., André, J. B., & Sperber, D. (2013). "A mutualistic approach to morality: The evolution of fairness by partner choice." *Behavioral and Brain Sciences, 36*(1), 59–78.
Baumard, N., & Boyer, P. (2013a). "Explaining moral religions." *Trends in Cognitive Sciences, 17*(6), 272–280.
Baumard, N., & Boyer, P. (2013b). "Religious beliefs as reflective elaborations on intuitions: A modified dual-process model." *Current Directions in Psychological Science, 22*(4), 295–300.

Baumard, N., & Chevallier, C. (2015). "The nature and dynamics of world religions: A life-history approach." *Proceedings of the Royal Society B, 282*, 20151593. https://doi.org/10.1098/rspb.2015.1593

Baumard, N., Hyafil, A., Morris, I., & Boyer, P. (2015). "Increased affluence explains the emergence of ascetic wisdoms and moralizing religions." *Current Biology, 25*(1), 10–15.

Beam, M. A., Hutchens, M. J., & Hmielowski, J. D. (2018). "Facebook news and (de)polarization: Reinforcing spirals in the 2016 US election." *Information, Communication and Society, 21*(7), 940–958.

Bekkouche, Y., & Cagé, J. (2018). *The price of a vote: Evidence from France, 1993–2014*. Retrieved from https://papers.ssrn.com/sol3/papers.cfm?abstract_id=3125220

Benjamin, L. T., & Simpson, J. A. (2009). The power of the situation: The impact of Milgram's obedience studies on personality and social psychology. *American Psychologist, 64*(1), 12–19.

Benjamin, P., & Shapiro, R. (1992). *The rational public: Fifty years of trends in Americans' policy preferences*. Chicago: University of Chicago Press.

Benkler, Y., Faris, R., & Roberts, H. (2018). *Network propaganda: Manipulation, disinformation, and radicalization in American politics*. New York: Oxford University Press.

Berelson, B. R., Lazarsfeld, P. F., McPhee, W. N., & McPhee, W. N. (1954). *Voting: A study of opinion formation in a presidential campaign*. Chicago: University of Chicago Press.

Berg, J., Dickhaut, J., & McCabe, K. (1995). "Trust, reciprocity, and social history." *Games and Economic Behavior, 10*(1), 122–142.

Bergstrom, B., & Boyer, P. (submitted). *Who mental systems believe: Effects of source on judgments of truth*.

Bergstrom, C. T., & Lachmann, M. (2001). "Alarm calls as costly signals of anti-predator vigilance: The watchful babbler game." *Animal Behaviour, 61*(3), 535–543.

Bernard, S., Proust, J., & Clément, F. (2015). "Four- to six-year-old children's sensitivity to reliability versus consensus in the endorsement of object labels." *Child Development, 86*(4), 1112–1124.

Berreman, G. D. (1971). "On the nature of caste in India: A review symposium on Louis Dumont's Homo Hierarchicus: 3 The Brahmannical View of Caste." *Contributions to Indian Sociology, 5*(1), 16–23.

Besley, T., & Burgess, R. (2002). "The political economy of government responsiveness: Theory and evidence from India." *Quarterly Journal of Economics, 117*(4), 1415–1451.

Birch, S. A., & Bloom, P. (2007). "The curse of knowledge in reasoning about false beliefs." *Psychological Science, 18*(5), 382–386.

Blake, T., Nosko, C., & Tadelis, S. (2015). "Consumer heterogeneity and paid search effectiveness: A large-scale field experiment." *Econometrica, 83*(1), 155–174.

Blakeslee, D. (2014). *Propaganda and politics in developing countries: Evidence from India.* Retrieved from https://papers.ssrn.com/sol3/papers.cfm?abstract_id=2542702

Blancke, S., Boudry, M., & Pigliucci, M. (2017). "Why do irrational beliefs mimic science? The cultural evolution of pseudoscience." *Theoria, 83*(1), 78–97.

Blumstein, D. T., Steinmetz, J., Armitage, K. B., & Daniel, J. C. (1997). "Alarm calling in yellow-bellied marmots: II. The importance of direct fitness." *Animal Behaviour, 53*(1), 173–184.

Boehm, C. (1999). *Hierarchy in the forest: The evolution of egalitarian behavior.* Cambridge, MA: Harvard University Press.

Bonaccio, S., & Dalal, R. S. (2006). "Advice taking and decision-making: An integrative literature review, and implications for the organizational sciences." *Organizational Behavior and Human Decision Processes, 101*(2), 127–151.

Bond, C. F. (2008). "Commentary: A few can catch a liar, sometimes: Comments on Ekman and O'Sullivan (1991), as well as Ekman, O'Sullivan, and Frank (1999)." *Applied Cognitive Psychology, 22*(9), 1298–1300.

Bond, C. F., & DePaulo, B. M. (2006). "Accuracy of deception judgments." *Personality and Social Psychology Review, 10*(3), 214–234.

Bond, C. F., Howard, A. R., Hutchison, J. L., & Masip, J. (2013). "Overlooking the obvious: Incentives to lie." *Basic and Applied Social Psychology, 35*(2), 212–221.

Bonnefon, J.-F., Hopfensitz, A., & De Neys, W. (2017). "Can we detect cooperators by looking at their face?" *Current Directions in Psychological Science, 26*(3), 276–281.

Borgia, G. (1985). "Bower quality, number of decorations and mating success of male satin bowerbirds (*Ptilonorhynchus violaceus*): An experimental analysis." *Animal Behaviour, 33*(1), 266–271.

Borgia, G. (1993). "The cost of display in the non-resource-based mating system of the satin bowerbird." *American Naturalist, 141*(5), 729–743.

Boss, L. P. (1997). "Epidemic hysteria: A review of the published literature." *Epidemiologic Reviews, 19*(2), 233–243.

Boxell, L., Gentzkow, M., & Shapiro, J. M. (2017). "Greater internet use is not associated with faster growth in political polarization among US demographic groups." *Proceedings of the National Academy of Sciences,* 201706588.

Boxell, L., Gentzkow, M., & Shapiro, J. M. (2018). "A note on internet use and the 2016 US presidential election outcome." *PloS One, 13*(7), e0199571.

Boyd, R., & Richerson, P. J. (1985). *Culture and the evolutionary process.* Chicago: University of Chicago Press.

Boyd, R., & Richerson, P. J. (2005). *The origin and evolution of cultures.* New York: Oxford University Press.

Boyer, P. (1994). *The naturalness of religious ideas: A cognitive theory of religion*. Los Angeles: University of California Press.

Boyer, P. (2001). *Religion explained*. London: Heinemann.

Boyer, P., & Baumard, N. (2018). "The diversity of religious systems across history." In J. R. Liddle & T. K. Shackelford (Eds.), *The Oxford handbook of evolutionary psychology and religion* (pp. 1–24). New York: Oxford University Press.

Boyer, P., & Parren, N. (2015). "Threat-related information suggests competence: A possible factor in the spread of rumors." *PloS One, 10*(6), e0128421.

Boyer, P., & Petersen, M. B. (2012). "The naturalness of (many) social institutions: Evolved cognition as their foundation." *Journal of Institutional Economics, 8*(1), 1–25.

Boyer, P., & Petersen, M. B. (2018). "Folk-economic beliefs: An evolutionary cognitive model." *Behavioral and Brain Sciences, 41*, e158.

Brain, P. (1986). *Galen on bloodletting: A study of the origins, development, and validity of his opinions, with a translation of the three works*. Cambridge: Cambridge University Press.

Brain, R. (1970). "Child-witches." In M. Douglas (Ed.), *Witchcraft confessions and accusations* (pp. 161–182). London: Routledge.

Brand, C. O., & Mesoudi, A. (2018). "Prestige and dominance based hierarchies exist in naturally occurring human groups, but are unrelated to task-specific knowledge." *Royal Society Open Science, 6*(6), 181621. https://doi.org/10.1098/rsos.181621

Brandenberger, D. (2012). *Propaganda state in crisis: Soviet ideology, indoctrination, and terror under Stalin, 1927–1941*. New Haven, CT: Yale University Press.

Braucher, J., & Orbach, B. (2015). "Scamming: The misunderstood confidence man." *Yale Journal of Law and the Humanities, 27*(2), 249–287.

Brennan, J. (2012). *The ethics of voting*. New York: Princeton University Press.

Brennan, J. (2016). *Against democracy*. Princeton, NJ: Princeton University Press.

Broockman, D. E., & Butler, D. M. (2017). "The causal effects of elite position-taking on voter attitudes: Field experiments with elite communication." *American Journal of Political Science, 61*(1), 208–221.

Broockman, D. E., & Green, D. P. (2014). "Do online advertisements increase political candidates' name recognition or favorability? Evidence from randomized field experiments." *Political Behavior, 36*(2), 263–289.

Brosseau-Liard, P. E., & Poulin-Dubois, D. (2014). "Sensitivity to confidence cues increases during the second year of life." *Infancy, 19*(5), 461–475.

Brown, R. (1965). *Social psychology*. New York: Free Press.

Buckner, H. T. (1965). "A theory of rumor transmission." *Public Opinion Quarterly, 29*(1), 54–70.

Burger, J. M., Girgis, Z. M., & Manning, C. C. (2011). "In their own words: Explaining obedience to authority through an examination of participants' comments." *Social Psychological and Personality Science, 2*(5), 460–466.

Burgess, T. H. (1839). *The physiology or mechanism of blushing*. London: Churchill.

Burns, J. (2012). "Race, diversity and pro-social behavior in a segmented society." *Journal of Economic Behavior and Organization, 81*(2), 366–378.

Burridge, K. O. L. (1972). "Tangu." In P. Lawrence & M. J. Meggitt (Eds.), *Gods, ghosts and men in Melanesia: Some religions of Australian New Guinea and the New Hebrides* (pp. 224–249). New York: Oxford University Press.

Bursztyn, L., Egorov, G., & Fiorin, S. (2019). *From extreme to mainstream: The erosion of social norms*. https://home.uchicago.edu/bursztyn/Bursztyn_Egorov_Fiorin_Extreme_Mainstream_2019_06_05.pdf.

Campagna, R. L., Mislin, A. A., Kong, D. T., & Bottom, W. P. (2016). "Strategic consequences of emotional misrepresentation in negotiation: The blowback effect." *Journal of Applied Psychology, 101*(5), 605–624.

Canetti, E. (1981). *Crowds and power* (C. Stewart, Trans.). New York: Noonday Press.

Caplow, T. (1947). "Rumors in war." *Social Forces, 25*(3), 298–302.

Carlsson, M., Dahl, G. B., & Rooth, D.-O. (2015). *Do politicians change public attitudes?* NBER Working Paper No. 21062. Retrieved from https://www.nber.org/papers/w21062

Caro, T. M. (1986a). "The functions of stotting: A review of the hypotheses." *Animal Behaviour, 34*(3), 649–662.

Caro, T. M. (1986b). "The functions of stotting in Thomson's gazelles: Some tests of the predictions." *Animal Behaviour, 34*(3), 663–684.

Carruthers, S. L. (2009). *Cold War captives: Imprisonment, escape, and brainwashing*. Los Angeles: University of California Press.

Castelain, T., Bernard, S., & Mercier, H. (2018). "Evidence that two-year-old children are sensitive to information presented in arguments." *Infancy, 23*(1), 124–135.

Castelain, T., Bernard, S., Van der Henst, J.-B., & Mercier, H. (2016). "The influence of power and reason on young Maya children's endorsement of testimony." *Developmental Science, 19*(6), 957–966.

Castelain, T., Girotto, V., Jamet, F., & Mercier, H. (2016). "Evidence for benefits of argumentation in a Mayan indigenous population." *Evolution and Human Behavior, 37*(5), 337–342.

Cavari, A., & Freedman, G. (2018). "Polarized mass or polarized few? Assessing the parallel rise of survey nonresponse and measures of polarization." *Journal of Politics, 80*(2), 719–725.

Chagnon, N. A. (1992). *Yanomamö: The fierce people* (4th ed.). New York: Holt, Rinehart and Winston.

Chan, M. S., Jones, C. R., Hall Jamieson, K., & Albarracin, D. (2017). Debunking: A meta-analysis of the psychological efficacy of messages countering misinformation. *Psychological Science, 28*(11), 1531–1546.

Chanel, O., Luchini, S., Massoni, S., & Vergnaud, J.-C. (2011). "Impact of information on intentions to vaccinate in a potential epidemic: Swine-origin influenza A (H1N1)." *Social Science and Medicine*, 72(2), 142–148.

Charness, G., & Dufwenberg, M. (2006). "Promises and partnership." *Econometrica*, 74(6), 1579–1601.

Cheatham, M. L. (2008). "The death of George Washington: An end to the controversy?" *American Surgeon*, 74(8), 770–774.

Chen, X., & Shi, T. (2001). "Media effects on political confidence and trust in the People's Republic of China in the post-Tiananmen period." *East Asia*, 19(3), 84–118.

Chen, Y., YeckehZaare, I., & Zhang, A. F. (2018). "Real or bogus: Predicting susceptibility to phishing with economic experiments." *PloS One*, 13(6), e0198213.

Cheney, D. L., & Seyfarth, R. M. (1988). "Assessment of meaning and the detection of unreliable signals by vervet monkeys." *Animal Behaviour*, 36(2), 477–486.

Cheney, D. L., & Seyfarth, R. M. (1990). *How monkeys see the world*. Chicago: University of Chicago Press.

Cheney, D. L., Seyfarth, R. M., & Silk, J. B. (1995). "The role of grunts in reconciling opponents and facilitating interactions among adult female baboons." *Animal Behaviour*, 50(1), 249–257.

Chiang, C.-F., & Knight, B. (2011). "Media bias and influence: Evidence from newspaper endorsements." *Review of Economic Studies*, 78(3), 795–820.

Chiarella, S. S., & Poulin-Dubois, D. (2013). "Cry babies and Pollyannas: Infants can detect unjustified emotional reactions." *Infancy*, 18(s1), E81–E96.

Chiarella, S. S., & Poulin-Dubois, D. (2015). "'Aren't you supposed to be sad?' Infants do not treat a stoic person as an unreliable emoter." *Infant Behavior and Development*, 38, 57–66.

Choleris, E., Guo, C., Liu, H., Mainardi, M., & Valsecchi, P. (1997). "The effect of demonstrator age and number on duration of socially-induced food preferences in house mouse (*Mus domesticus*)." *Behavioural Processes*, 41(1), 69–77.

Chorus, A. (1953). "The basic law of rumor." *Journal of Abnormal and Social Psychology*, 48(2), 313–314.

Chwe, M. (2001). *Rational ritual*. New York: Princeton University Press.

Claidière, N., Trouche, E., & Mercier, H. (2017). "Argumentation and the diffusion of counter-intuitive beliefs." *Journal of Experimental Psychology: General*, 146(7), 1052–1066.

Clauset, A., Arbesman, S., & Larremore, D. B. (2015). "Systematic inequality and hierarchy in faculty hiring networks." *Science Advances*, 1(1), e1400005.

Clément, F. (2006). *Les mécanismes de la crédulité*. Geneva: Librairie Droz.

Clément, F. (2010). "To trust or not to trust? Children's social epistemology." *Review of Philosophy and Psychology*, 1(4), 1–19.

Clément, F., Koenig, M. A., & Harris, P. (2004). "The ontogenesis of trust." *Mind and Language*, 19(4), 360–379.

Cohen, I. B. (1985). *Revolution in science*. Cambridge, MA: Harvard University Press.

Cohn, N. (1970). *The pursuit of the millennium*. St. Albans: Paladin.

Collins, P. J., Hahn, U., von Gerber, Y., & Olsson, E. J. (2018). "The bi-directional relationship between source characteristics and message content." *Frontiers in Psychology*, 9. Retrieved from https://www.frontiersin.org/articles/10.3389/fpsyg.2018.00018/full

Condorcet, J. A. N. (1785). *Essai sur l'application de l'analyse à la probabilité des décisions rendues à la pluralité des voix*.

Condorcet, J. A. N. (1797). *Esquisse d'un tableau historique des progrès de l'esprit humain*.

Conner, A. W. (2000). "True confessions? Chinese confessions then and now." In K. G. Turner, J. V. Feinerman, & R. K. Guy (Eds.), *The limits of the rule of law in China* (pp. 132–162). Seattle: University of Washington Press.

Conradt, L., & List, C. (2009). "Group decisions in humans and animals: A survey." *Philosophical Transactions of the Royal Society of London B: Biological Sciences*, 364(1518), 719–742.

Conradt, L., & Roper, T. J. (2003). "Group decision-making in animals." *Nature*, 421(6919), 155–158.

Corriveau, K. H., & Harris, P. L. (2010). "Preschoolers (sometimes) defer to the majority in making simple perceptual judgments." *Developmental Psychology*, 46(2), 437–445.

Costas-Pérez, E., Solé-Ollé, A., & Sorribas-Navarro, P. (2012). "Corruption scandals, voter information, and accountability." *European Journal of Political Economy*, 28(4), 469–484.

Couch, C. J. (1968). "Collective behavior: An examination of some stereotypes." *Social Problems*, 15(3), 310–322.

Couillard, N. L., & Woodward, A. L. (1999). "Children's comprehension of deceptive points." *British Journal of Developmental Psychology*, 17(4), 515–521.

Coviello, L., Sohn, Y., Kramer, A. D., Marlow, C., Franceschetti, M., Christakis, N. A., & Fowler, J. H. (2014). "Detecting emotional contagion in massive social networks." *PloS One*, 9(3), e90315.

Crivelli, C., & Fridlund, A. J. (2018). "Facial displays are tools for social influence." *Trends in Cognitive Sciences*, 22(5), 388–399.

Crowell, A., & Kuhn, D. (2014). "Developing dialogic argumentation skills: A 3-year intervention study." *Journal of Cognition and Development*, 15(2), 363–381.

Davies, S. R. (1997). *Popular opinion in Stalin's Russia: Terror, propaganda and dissent, 1934–1941*. Cambridge: Cambridge University Press.

Dawkins, R. (2010). *A devil's chaplain: Selected writings*. London: Hachette UK.

Dawkins, R., & Krebs, J. R. (1978). "Animal signals: Information or manipulation?" In J. R. Krebs & N. B. Davies (Eds.), *Behavioural ecology: An evolutionary approach* (pp. 282–309). Oxford: Basil Blackwell Scientific Publications.

Dediu, D., & Levinson, S. C. (2018). "Neanderthal language revisited: Not only us." *Current Opinion in Behavioral Sciences, 21,* 49–55.

Dehaene, S. (1999). *The number sense: How the mind creates mathematics.* Oxford: Oxford University Press.

DellaVigna, S., & Gentzkow, M. (2010). "Persuasion: Empirical evidence." *Annual Review of Economics, 2*(1), 643–669.

DellaVigna, S., & Kaplan, E. (2007). "The Fox News effect: Media bias and voting." *Quarterly Journal of Economics, 122*(3), 1187–1234.

Delton, A. W., & Cimino, A. (2010). "Exploring the evolved concept of NEWCOMER: Experimental tests of a cognitive model." *Evolutionary Psychology, 8*(2), 147470491000800220.

Delumeau, J. (1977). *Catholicism between Luther and Voltaire.* Philadelphia: Westminster Press.

Del Vicario, M., Scala, A., Caldarelli, G., Stanley, H. E., & Quattrociocchi, W. (2017). "Modeling confirmation bias and polarization." *Scientific Reports, 7,* 40391.

Demick, B. (2010). *Nothing to envy: Real lives in North Korea.* New York: Spiegel and Grau.

Dennett, D. C. (1995). *Darwin's dangerous idea.* London: Penguin Books.

DePaulo, B. M. (1992). "Nonverbal behavior and self-presentation." *Psychological Bulletin, 111*(2), 203–243.

DePaulo, B. M., Kashy, D. A., Kirkendol, S. E., Wyer, M. M., & Epstein, J. A. (1996). "Lying in everyday life." *Journal of Personality and Social Psychology, 70*(5), 979–995.

DePaulo, B. M., Lindsay, J. J., Malone, B. E., Muhlenbruck, L., Charlton, K., & Cooper, H. (2003). "Cues to deception." *Psychological Bulletin, 129*(1), 74–118.

Desmet, K., & Wacziarg, R. (2018). *The cultural divide.* NBER Working Paper No. 24630. Retrived from https://www.nber.org/papers/w24630

Desrochers, S., Morissette, P., & Ricard, M. (1995). "Two perspectives on pointing in infancy." In C. Moore & P. Dunham (Eds.), *Joint attention: Its origins and role in development* (pp. 85–101). Hillsdale, NJ: Erlbaum.

Dessalles, J.-L. (2007). *Why we talk: The evolutionary origins of language.* Cambridge: Oxford University Press.

Deutsch, M., & Gerard, H. B. (1955). "A study of normative and informational social influences upon individual judgment." *Journal of Abnormal and Social Psychology, 51*(3), 629–636.

De Vries, C. E., Hobolt, S. B., & Tilley, J. (2018). "Facing up to the facts: What causes economic perceptions?" *Electoral Studies, 51,* 115–122.

de Waal, F. B. M. (1982). *Chimpanzee politics*. New York: Harper and Row.
Dewitt, S. H., Lagnado, D., & Fenton, N. E. (submitted). *Updating prior beliefs based on ambiguous evidence*. Retrieved from https://www.researchgate.net/publication/326610460_Updating_Prior_Beliefs_Based_on_Ambiguous_Evidence
Dezecache, G. (2015). "Human collective reactions to threat." *Wiley Interdisciplinary Reviews: Cognitive Science*, 6(3), 209–219.
Dezecache, G., Martin, J. R., Tessier, C., Safra, L., Pitron, V., Nuss, P., & Grèzes, J. (submitted). *Social strategies in response to deadly danger during a mass shooting*.
Dezecache, G., Mercier, H., & Scott-Phillips, T. C. (2013). "An evolutionary approach to emotional communication." *Journal of Pragmatics*, 59(B), 221–233.
Dickson, G. (2007). *The Children's Crusade: Medieval history, modern mythistory*. London: Palgrave Macmillan.
DiFonzo, N. (2010). "Ferreting facts or fashioning fallacies? Factors in rumor accuracy." *Social and Personality Psychology Compass*, 4(11), 1124–1137.
DiFonzo, N., & Bordia, P. (2007). *Rumor psychology: Social and organizational approaches*. Washington, DC: American Psychological Association.
Diggory, J. C. (1956). "Some consequences of proximity to a disease threat." *Sociometry*, 19(1), 47–53.
Dikötter, F. (2010). *Mao's great famine: The history of China's most devastating catastrophe, 1958–1962*. New York: Walker and Company.
Dimberg, U., Thunberg, M., & Elmehed, K. (2000). "Unconscious facial reactions to emotional facial expressions." *Psychological Science*, 11(1), 86–89.
Dixon, G., Hmielowski, J., & Ma, Y. (2017). "Improving climate change acceptance among US conservatives through value-based message targeting." *Science Communication*, 39(4), 520–534.
Dockendorff, M., & Mercier, H. (in preparation). *Argument transmission as the weak link in the correction of political misbeliefs*.
Donovan, P. (2004). *No way of knowing: Crime, urban legends and the internet*. London: Routledge.
Drizin, S. A., & Leo, R. A. (2003). "The problem of false confessions in the post-DNA world." *North Carolina Law Review*, 82, 891–1007.
Druckman, J. N., Levendusky, M. S., & McLain, A. (2018). "No need to watch: How the effects of partisan media can spread via interpersonal discussions." *American Journal of Political Science*, 62(1), 99–112.
Drury, J., Novelli, D., & Stott, C. (2013). "Psychological disaster myths in the perception and management of mass emergencies." *Journal of Applied Social Psychology*, 43(11), 2259–2270.
Dubois, E., & Blank, G. (2018). "The echo chamber is overstated: The moderating effect of political interest and diverse media." *Information, Communication and Society*, 21(5), 729–745.

Dubreuil, B. (2010). "Paleolithic public goods games: Why human culture and cooperation did not evolve in one step." *Biology and Philosophy*, 25(1), 53–73.

Dumont, L. (1980). *Homo hierarchicus: The caste system and its implications.* Chicago: University of Chicago Press.

Dunlap, A. S., Nielsen, M. E., Dornhaus, A., & Papaj, D. R. (2016). "Foraging bumble bees weigh the reliability of personal and social information." *Current Biology*, 26(9), 1195–1199.

Durante, R., & Gutierrez, E. (2014). *Political advertising and voting intentions: Evidence from exogenous variation in ads viewership.* Unpublished manuscript. Retrieved from https://spire.sciencespo.fr/hdl:/2441/26lctatf2u8130f8nkn7j2230h/resources/wp-mexico-political-advertising.pdf.

Durbach, N. (2000). "'They might as well brand us': Working-class resistance to compulsory vaccination in Victorian England." *Social History of Medicine*, 13(1), 45–63.

Ebrahim, G. J. (1968). "Mass hysteria in school children: Notes on three outbreaks in East Africa." *Clinical Pediatrics*, 7(7), 437–438.

Ecker, U. K., O'Reilly, Z., Reid, J. S., & Chang, E. P. (2019). The effectiveness of short-format refutational fact-checks. *British Journal of Psychology.* https://doi.org/10.1111/bjop.12383.

Edwards, K., & Smith, E. E. (1996). "A disconfirmation bias in the evaluation of arguments." *Journal of Personality and Social Psychology*, 71(1), 5–24.

Einav, S. (2017). "Thinking for themselves? The effect of informant independence on children's endorsement of testimony from a consensus." *Social Development*, 27(1), 73–86.

Einav, S., & Robinson, E. J. (2011). "When being right is not enough: Four-year-olds distinguish knowledgeable informants from merely accurate informants." *Psychological Science*, 22(10), 1250–1253.

Ekelund, R. B., Tollison, R. D., Anderson, G. M., Hébert, R. F., & Davidson, A. B. (1996). *Sacred trust: The medieval church as an economic firm.* New York: Oxford University Press.

Ekman, P. (2001). *Telling lies: Clues to deceit in the marketplace, politics, and marriage.* New York: Norton

Ekman, P. (2009). "Lie catching and microexpressions." In C. Martin (Ed.), *The philosophy of deception* (pp.118–133). Oxford: Oxford University Press.

Enders, A. M., & Armaly, M. T. (2018). "The differential effects of actual and perceived polarization." *Political Behavior*, https://doi.org/10.1007/s11109-018-9476-2.

Engelberg, J. W., & Gouzoules, H. (2019). "The credibility of acted screams: Implications for emotional communication research." *Quarterly Journal of Experimental Psychology*, 72(8), 1889–1902.

Epler, D. C. (1980). "Bloodletting in early Chinese medicine and its relation to the origin of acupuncture." *Bulletin of the History of Medicine*, 54(3), 337–367.

Eriksson, K. (2012). "The nonsense math effect." *Judgment and Decision Making, 7*(6), 746–749.

Estlund, D. (1994). "Opinion leaders, independence, and Condorcet's jury theorem." *Theory and Decision, 36*(2), 131–162.

Evans, H., & Bartholomew, R. (2009). *Outbreak! The encyclopedia of extraordinary social behavior.* New York: Anomalist Books.

Evans-Pritchard, E. E. (1937). *Witchcraft, magic and oracles among the Azande.* Retrieved from eHRAF: World Cultures database.

Faasse, K., Chatman, C. J., & Martin, L. R. (2016). "A comparison of language use in pro- and anti-vaccination comments in response to a high profile Facebook post." *Vaccine, 34*(47), 5808–5814.

Facchini, G., Margalit, Y., & Nakata, H. (2016). *Countering public opposition to immigration: The impact of information campaigns.* Unpublished article. Retrieved from https://papers.ssrn.com/sol3/papers.cfm?abstract_id=2887349.

Fadda, M., Allam, A., & Schulz, P. J. (2015). "Arguments and sources on Italian online forums on childhood vaccinations: Results of a content analysis." *Vaccine, 33*(51), 7152–7159.

Fershtman, C., & Gneezy, U. (2001). "Discrimination in a segmented society: An experimental approach." *Quarterly Journal of Economics, 116*(1), 351–377.

Fifield, A. (2019). *The Great Successor: The divinely perfect destiny of brilliant comrade Kim Jong Un.* New York: PublicAffairs.

Fiorina, M. P., Abrams, S. J., & Pope, J. (2005). *Culture war? The myth of a polarized America.* New York: Pearson Longman.

FitzGibbon, C. D., & Fanshawe, J. H. (1988). "Stotting in Thomson's gazelles: An honest signal of condition." *Behavioral Ecology and Sociobiology, 23*(2), 69–74.

Flaxman, S., Goel, S., & Rao, J. M. (2016). "Filter bubbles, echo chambers, and online news consumption." *Public Opinion Quarterly, 80*(S1), 298–320.

Fletcher, R., & Nielsen, R. K. (2017). "Are news audiences increasingly fragmented? A cross-national comparative analysis of cross-platform news audience fragmentation and duplication." *Journal of Communication, 67*(4), 476–498.

Foddy, M., Platow, M. J., & Yamagishi, T. (2009). "Group-based trust in strangers: The role of stereotypes and expectations." *Psychological Science, 20*(4), 419–422.

Fodor, J. (1983). *The modularity of mind.* Cambridge, MA: MIT Press.

Foerster, A., Wirth, R., Herbort, O., Kunde, W., & Pfister, R. (2017). "Lying upside-down: Alibis reverse cognitive burdens of dishonesty." *Journal of Experimental Psychology: Applied, 23*(3), 301–319.

Fourney, A., Racz, M. Z., Ranade, G., Mobius, M., & Horvitz, E. (2017). "Geographic and temporal trends in fake news consumption during the 2016 US presidential election." *Proceedings of the 2017 ACM Conference on Information and Knowledge Management,* 2071–2074.

Frank, R. H. (1988). *Passions within reason: The strategic role of emotions.* New York: Norton.

Frederick, S. (2005). "Cognitive reflection and decision making." *Journal of Economic Perspectives, 19*(4), 25–42.

Fresco, N. (1980). "Les redresseurs de morts. Chambres à gaz: la bonne nouvelle." Comment on révise l'histoire. *Les Temps Modernes, 407,* 2150–2211.

Freud, S. (1905). "Fragment of an analysis of a case of hysteria." In E. Jones (Ed.), *Collected papers* (pp. 13–146). New York: Basic Books.

Friend, R., Rafferty, Y., & Bramel, D. (1990). "A puzzling misinterpretation of the Asch 'conformity' study." *European Journal of Social Psychology, 20*(1), 29–44.

Fusaroli, R., Bahrami, B., Olsen, K., Roepstorff, A., Rees, G., Frith, C., & Tylén, K. (2012). "Coming to terms quantifying the benefits of linguistic coordination." *Psychological Science,* 0956797612436816.

Futaba, I., & McCormack, G. (1984). "Crime, confession and control in contemporary Japan." *Law Context: A Socio-Legal Journal, 2,* 1–30.

Galler, J. S. (2007). *Logic and argumentation in "The Book of Concord"* (unpublished doctoral dissertation). University of Texas at Austin.

Gallup, A. C., Chong, A., & Couzin, I. D. (2012). "The directional flow of visual information transfer between pedestrians." *Biology Letters, 8*(4), 520–522.

Gallup, A. C., Hale, J. J., Sumpter, D. J., Garnier, S., Kacelnik, A., Krebs, J. R., & Couzin, I. D. (2012). "Visual attention and the acquisition of information in human crowds." *Proceedings of the National Academy of Sciences, 109*(19), 7245–7250.

Galton, F. (1907). "Vox populi." *Nature, 75*(7), 450–451.

Gamson, W. A. (1992). *Talking politics.* Cambridge: Cambridge University Press.

Gang, Q., & Bandurski, D. (2011). "China's emerging public sphere: The impact of media commercialization, professionalism, and the internet in an era of transition." In S. L. Shirk (Ed.), *Changing media, changing China* (pp. 38–76). New York: Oxford University Press.

Garcia, J., Kimeldorf, D. J., & Koelling, R. A. (1955). "Conditioned aversion to saccharin resulting from exposure to gamma radiation." *Science, 122*(3160), 157–158.

Gelman, A., Goel, S., Rivers, D., & Rothschild, D. (2016). "The mythical swing voter." *Quarterly Journal of Political Science, 11*(1), 103–130.

Gelman, A., & King, G. (1993). "Why are American presidential election campaign polls so variable when votes are so predictable?" *British Journal of Political Science, 23*(4), 409–451.

Gendelman, M. (2013). *A tale of two soldiers: The unexpected friendship between a WWII American Jewish sniper and a German military pilot.* Minneapolis, MN: Hillcrest Publishing Group.

Genovese, E. D. (1974). *Roll, Jordan, roll: The world the slaves made.* New York: Pantheon.

Gentzkow, M., & Shapiro, J. M. (2011). "Ideological segregation online and offline." *Quarterly Journal of Economics*, 126(4), 1799–1839.

Gervais, W. M., & Norenzayan, A. (2012). "Analytic thinking promotes religious disbelief." *Science*, 336(6080), 493–496.

Gervais, W. M., van Elk, M., Xygalatas, D., McKay, R. T., Aveyard, M., Buchtel, E. E., . . . Riekki, T. (2018). "Analytic atheism: A cross-culturally weak and fickle phenomenon?" *Judgment and Decision Making*, 13(3), 268–274.

Gilbert, D. T. (1991). "How mental systems believe." *American Psychologist*, 46(2), 107–119.

Gilbert, D. T., Krull, D. S., & Malone, P. S. (1990). "Unbelieving the unbelievable: Some problems in the rejection of false information." *Journal of Personality and Social Psychology*, 59(4), 601–613.

Gilbert, D. T., Tafarodi, R. W., & Malone, P. S. (1993). "You can't not believe everything you read." *Journal of Personality and Social Psychology*, 65(2), 221–233.

Gilsenan, M. (1976). "Lying, honor, and contradiction." In B. Kapferer (Ed.), *Transaction and Meaning: Directions in the Anthropology of Exchange and Symbolic Behavior* (pp. 191–219). Philadelphia: Institute for the Study of Human Issues.

Gino, F. (2008). "Do we listen to advice just because we paid for it? The impact of advice cost on its use." *Organizational Behavior and Human Decision Processes*, 107(2), 234–245.

Ginzburg, C. (2013). *The cheese and the worms: The cosmos of a sixteenth-century miller.* Baltimore: Johns Hopkins University Press.

Glaeser, E. L., Laibson, D. I., Scheinkman, J. A., & Soutter, C. L. (2000). "Measuring trust." *Quarterly Journal of Economics*, 115(3), 811–846.

Goldacre, B. (2014). *Bad pharma: How drug companies mislead doctors and harm patients.* London: Macmillan.

Goldman, A. I. (2001). "Experts: Which ones should you trust?" *Philosophy and Phenomenological Research*, 63(1), 85–110.

Goues, C. L., Brun, Y., Apel, S., Berger, E., Khurshid, S., & Smaragdakis, Y. (2017). *Effectiveness of anonymization in double-blind review.* Retrieved from https://arxiv.org/abs/1709.01609

Gould, J. L., & Gould, C. G. (1982). "The insect mind: Physics or metaphysics?" In D. R. Griffin (Ed.), *Animal mind—Human mind* (pp. 269–298). Berlin: Springer-Verlag.

Granovetter, M. (1978). "Threshold models of collective behavior." *American Journal of Sociology*, 83(6), 1420–1443.

Greene, E. D. (1990). "The logic of university students' misunderstanding of natural selection." *Journal of Research in Science Teaching*, 27(9), 875–885.

Greenspan, S. (2008). *Annals of gullibility: Why we get duped and how to avoid it.* New York: ABC-CLIO.

Greenwald, A. G. (1968). "Cognitive learning, cognitive response to persuasion, and attitude change." In A. G. Greenwald, T. C. Brock, & T. M. Ostrom (Eds.), *Psychological foundations of attitudes* (pp. 147–170). New York: Academic Press.

Griggs, R. A. (2015). "The disappearance of independence in textbook coverage of Asch's social pressure experiments." *Teaching of Psychology, 42*(2), 137–142.

Griggs, R. A., & Whitehead, G. I. (2015). "Coverage of Milgram's obedience experiments in social psychology textbooks: Where have all the criticisms gone?" *Teaching of Psychology, 42*(4), 315 322.

Grigorieff, A., Roth, C., & Ubfal, D. (2018). "Does information change attitudes towards immigrants? Representative evidence from survey experiments." Unpublished article. Retrieved from https://papers.ssrn.com/sol3/papers.cfm?abstract_id=2768187.

Grinberg, N., Joseph, K., Friedland, L., Swire-Thompson, B., & Lazer, D. (2019). "Fake news on Twitter during the 2016 US presidential election." *Science, 363*(6425), 374–378.

Gross, D. K. (2018). *Documents of the Salem witch trials.* Santa Barbara, CA: ABC-CLIO.

Grove, M. (2018). "Strong conformity requires a greater proportion of asocial learning and achieves lower fitness than a payoff-based equivalent." *Adaptive Behavior, 26*(6), 323–333.

Gudjonsson, G. H. (2003). *The psychology of interrogations and confessions: A handbook.* New York: Wiley.

Gudjonsson, G. H., & Sigurdsson, J. F. (1994). "How frequently do false confessions occur? An empirical study among prison inmates." *Psychology, Crime and Law, 1*(1), 21–26.

Gudjonsson, G. H., Sigurdsson, J. F., Bragason, O. O., Einarsson, E., & Valdimarsdottir, E. B. (2004). "Confessions and denials and the relationship with personality." *Legal and Criminological Psychology, 9*(1), 121–133.

Gudjonsson, G. H., Sigurdsson, J. F., & Einarsson, E. (2004). "The role of personality in relation to confessions and denials." *Psychology, Crime and Law, 10*(2), 125–135.

Guess, A. (2016). *Media choice and moderation: Evidence from online tracking data.* Unpublished manuscript, New York University.

Guess, A., & Coppock, A. (2015). *Back to Bayes: Confronting the evidence on attitude polarization.* Unpublished manuscript. Retrieved from https://pdfs.semanticscholar.org/23fc/c2e9e5706a766148e71624dc0f78e3cbf8ef.pdf.

Guess, A., & Coppock, A. (2018). "Does counter-attitudinal information cause backlash? Results from three large survey experiments." *British Journal of Political Science.* https://doi.org/10.1017/S0007123418000327.

Guess, A., Nagler, J., & Tucker, J. (2019). "Less than you think: Prevalence and predictors of fake news dissemination on Facebook." *Science Advances*, 5(1), eaau4586.

Guess, A., Nyhan, B., & Reifler, J. (2018). *Selective exposure to misinformation: Evidence from the consumption of fake news during the 2016 US presidential campaign*. Retrieved from http://www.ask-force.org/web/Fundamentalists/Guess-Selective-Exposure-to-Misinformation-Evidence-Presidential-Campaign-2018.pdf.

Gupta, G., Mahmud, M., Maitra, P., Mitra, S., & Neelim, A. (2013). *Religion, minority status and trust: Evidence from a field experiment*. Retrieved from https://www.researchgate.net/profile/Minhaj_Mahmud2/publication/313006388_Religion_Minority_Status_and_Trust_Evidence_from_a_Field_Experiment/links/588c2e7daca272fa50ddeoa6/Religion-Minority-Status-and-Trust-Evidence-from-a-Field-Experiment.pdf.

Hahn, U., & Oaksford, M. (2007). "The rationality of informal argumentation: A Bayesian approach to reasoning fallacies." *Psychological Review*, 114(3), 704–732.

Haig, D. (1993). "Genetic conflicts in human pregnancy." *Quarterly Review of Biology*, 68(4), 495–532.

Haig, D. (1996). "Placental hormones, genomic imprinting, and maternal-fetal communication." *Journal of Evolutionary Biology*, 9(3), 357–380.

Hall, J. R. (2009). "Apocalyptic and millenarian movements." In D. A. Snow, D. della Porta, B. Klandermans, & D. McAdam (Eds.), *The Wiley-Blackwell encyclopedia of social and political movements* (pp. 1–3). London: Wiley-Blackwell.

Hall, J. R. (2013). *Apocalypse: From antiquity to the empire of modernity*. Indianapolis: Wiley.

Han, S. (2018). Neurocognitive basis of racial ingroup bias in empathy. *Trends in Cognitive Sciences*, 2(5), 400–421.

Harding, H. (1993). "The Chinese state in crisis, 1966–9." In R. MacFarquhar (Ed.), *The politics of China, 1949–1989* (pp. 148–247). New York: Cambridge University Press.

Harper, E. B. (1968). "Social consequences of an unsuccessful low caste movement." In J. Silverberg (Ed.), *Social Mobility in the Caste System in India* (pp. 36–65). The Hague: Mouton.

Harris, P. L. (2012). *Trusting what you're told: How children learn from others*. Cambridge, MA: Belknap Press of Harvard University Press.

Harris, P. L., Koenig, M. A., Corriveau, K. H., & Jaswal, V. K. (2018). "Cognitive foundations of learning from testimony." *Annual Review of Psychology*, 69(1), 251–273.

Hartwig, M., & Bond, C. H. (2011). "Why do lie-catchers fail? A lens model meta-analysis of human lie judgments." *Psychological Bulletin*, 137(4), 643–659.

Haslam, N., Loughnan, S., & Perry, G. (2014). "Meta-Milgram: An empirical synthesis of the obedience experiments." *PloS One, 9*(4), e93927.

Hassig, R., & Oh, K. (2009). *The hidden people of North Korea: Everyday life in the hermit kingdom*. London: Rowman and Littlefield.

Hasson, O. (1991). "Pursuit-deterrent signals: Communication between prey and predator." *Trends in Ecology and Evolution, 6*(10), 325–329.

Hasson, U., Simmons, J. P., & Todorov, A. (2005). "Believe it or not: On the possibility of suspending belief." *Psychological Science, 16*(7), 566–571.

Hastie, R., & Kameda, T. (2005). "The robust beauty of majority rules in group decisions." *Psychological Review, 112*(2), 494–508.

Hatfield, E., Cacioppo, J. T., & Rapson, R. L. (1994). *Emotional contagion*. Cambridge: Cambridge University Press.

Haun, D. B. M., & Tomasello, M. (2011). "Conformity to peer pressure in preschool children." *Child Development, 82*(6), 1759–1767.

Heathers, J. A., Fayn, K., Silvia, P. J., Tiliopoulos, N., & Goodwin, M. S. (2018). "The voluntary control of piloerection." *PeerJ Preprints, 6*, e26594v1.

Heckewelder, J.G.E. (1876). *History, manners, and customs of the Indian nations: Who once inhabited Pennsylvania and the neighboring states*. Philadelphia: Historical Society of Pennsylvania.

Henrich, J. (2009). "The evolution of costly displays, cooperation and religion: Credibility enhancing displays and their implications for cultural evolution." *Evolution and Human Behavior, 30*(4), 244–260.

Henrich, J. (2015). *The secret of our success: How culture is driving human evolution, domesticating our species, and making us smarter*. Princeton, NJ: Princeton University Press.

Henrich, J., & Boyd, R. (1998). "The evolution of conformist transmission and the emergence of between-group differences." *Evolution and Human Behavior, 19*(4), 215–241.

Henrich, J., & Broesch, J. (2011). "On the nature of cultural transmission networks: Evidence from Fijian villages for adaptive learning biases." *Philosophical Transactions of the Royal Society of London B: Biological Sciences, 366*(1567), 1139–1148.

Henrich, J., & Gil-White, F. J. (2001). "The evolution of prestige: Freely conferred deference as a mechanism for enhancing the benefits of cultural transmission." *Evolution and Human Behavior, 22*(3), 165–196.

Hepach, R., Vaish, A., & Tomasello, M. (2013). "Young children sympathize less in response to unjustified emotional distress." *Developmental Psychology, 49*(6), 1132–1138.

Heraclitus. (2001). *Fragments: The collected wisdom of Heraclitus* (B. Haxton, Trans.). London: Viking Adult.

Herley, C. (2012). "Why do Nigerian scammers say they are from Nigeria?" *WEIS*. Retrieved from http://infosecon.net/workshop/downloads/2012/pdf/Why_do _Nigerian_Scammers_Say_They_are_From_Nigeria.pdf.

Hernon, I. (2006). *Riot! Civil insurrection from Peterloo to the present day.* New York: Pluto Press.

Higham, J. P. (2013). "How does honest costly signaling work?" *Behavioral Ecology*, 25(1), 8–11.

Hill, K., & Kintigh, K. (2009). "Can anthropologists distinguish good and poor hunters? Implications for hunting hypotheses, sharing conventions, and cultural transmission." *Current Anthropology*, 50(3), 369–378.

Hill, S. J. (2017). "Learning together slowly: Bayesian learning about political facts." *Journal of Politics*, 79(4), 1403–1418.

Hill, S. J., Lo, J., Vavreck, L., & Zaller, J. (2013). "How quickly we forget: The duration of persuasion effects from mass communication." *Political Communication*, 30(4), 521–547.

Hirschfeld, L. A. (2002). "Why don't anthropologists like children?" *American Anthropologist*, 104(2), 611–627.

Hitler, A. (1939). *Mein Kampf* (J. Murphy, Trans.). London: Hurst and Blackett.

Hoare, G., & Sperber, N. (2015). *An introduction to Antonio Gramsci: His life, thought and legacy.* London: Bloomsbury.

Hodges, B. H., & Geyer, A. L. (2006). "A nonconformist account of the Asch experiments: Values, pragmatics, and moral dilemmas." *Personality and Social Psychology Review*, 10(1), 2–19.

Hoffmann, D. L., Standish, C. D., García-Diez, M., Pettitt, P. B., Milton, J. A., Zilhão, J., ... De Balbín, R. (2018). "U-Th dating of carbonate crusts reveals Neandertal origin of Iberian cave art." *Science*, 359(6378), 912–915.

Hofman, D., Bos, P. A., Schutter, D. J., & van Honk, J. (2012). "Fairness modulates nonconscious facial mimicry in women." *Proceedings of the Royal Society of London B: Biological Sciences*, 279(1742), 3535–3539.

Holbach, P.H.T.B.d'. (1835). *Christianity unveiled: Being an examination of the principles and effects of the Christian religion.* New York: Johnson.

Hollander, M. M., & Turowetz, J. (2017). "Normalizing trust: Participants' immediately post-hoc explanations of behaviour in Milgram's "obedience" experiments." *British Journal of Social Psychology*, 56(4), 655–674.

Honts, C. R., & Hartwig, M. (2014). "Credibility assessment at portals." In D. C. Raskin, C. R. Honts, & J. C. Kircher (Eds.), *Credibility assessment* (pp. 37–61). Amsterdam: Elsevier.

Hopkins, D. J., & Ladd, J. M. (2014). "The consequences of broader media choice: Evidence from the expansion of Fox News." *Quarterly Journal of Political Science*, 9(1), 115–135.

Hopkins, D. J., Sides, J., & Citrin, J. (2019). "The muted consequences of correct information about immigration." *Journal of Politics, 81*(1), 315–320.

Horne, Z., Powell, D., Hummel, J. E., & Holyoak, K. J. (2015). "Countering antivaccination attitudes." *Proceedings of the National Academy of Sciences, 112*(33), 10321–10324.

Horowitz, D. L. (2001). *The deadly ethnic riot.* Berkeley: University of California Press.

Hovland, C. I. (1954). "The effects of the mass media of communication." In L. Gardner (Ed.), *Handbook of social psychology* (pp. 244–252). Cambridge MA: Addison-Wesley.

Howard, G. (1983). *Frames of mind: The theory of multiple intelligences.* New York: Basic Books.

Hu, J., Whalen, A., Buchsbaum, D., Griffiths, T., & Xu, F. (2015). "Can children balance the size of a majority with the quality of their information?" *Proceedings of the Cognitive Science Society Conference.* Pasadena, California, July 22–25.

Hu, Y., Lodish, L. M., & Krieger, A. M. (2007). "An analysis of real world TV advertising tests: A 15-year update." *Journal of Advertising Research, 47*(3), 341–353.

Huang, H. (2017). "A war of (mis)information: The political effects of rumors and rumor rebuttals in an authoritarian country." *British Journal of Political Science, 47*(2), 283–311.

Huckfeldt, R. (2001). "The social communication of political expertise." *American Journal of Political Science, 45*(2), 425–438.

Huckfeldt, R., Pietryka, M. T., & Reilly, J. (2014). "Noise, bias, and expertise in political communication networks." *Social Networks, 36,* 110–121.

Hutton, R. (2017). *The witch: A history of fear, from ancient times to the present.* New Haven, CT: Yale University Press.

Iannaccone, L. R. (1992). "Sacrifice and stigma: Reducing free-riding in cults, communes, and other collectives." *Journal of Political Economy, 100*(2), 271–291.

Iannaccone, L. R. (2006). "The market for martyrs." *Interdisciplinary Journal of Research on Religion, 2*(4), 1–28.

Inbau, F., Reid, J., Buckley, J., & Jayne, B. (2001). *Criminal interrogation and confessions* (4th ed.). Gaithersberg, MD: Aspen.

Isberner, M.-B., & Richter, T. (2013). "Can readers ignore implausibility? Evidence for nonstrategic monitoring of event-based plausibility in language comprehension." *Acta Psychologica, 142*(1), 15–22.

Isberner, M.-B., & Richter, T. (2014). "Does validation during language comprehension depend on an evaluative mindset?" *Discourse Processes, 51*(1–2), 7–25.

Isenberg, D. J. (1986). "Group polarization: A critical review and meta-analysis." *Journal of Personality and Social Psychology, 50*(6), 1141–1151.

Iyengar, S., & Kinder, D. R. (1987). *News that matters: Television and public opinion.* Chicago: University of Chicago Press.

Iyengar, S., Lelkes, Y., Levendusky, M., Malhotra, N., & Westwood, S. J. (2019). "The origins and consequences of affective polarization in the United States." *Annual Review of Political Science, 22,* 129–146.

Janis, I. L. (1951). *Air war and emotional stress: Psychological studies of bombing and civilian defense.* New York: McGraw-Hill.

Jeffries, S. (2016). *Grand hotel abyss: The lives of the Frankfurt School.* New York: Verso.

Jiménez, Á. V., & Mesoudi, A. (2019). "Prestige-biased social learning: Current evidence and outstanding questions." *Palgrave Communications, 5*(1), 20. Retrieved from https://www.nature.com/articles/s41599-019-0228-7

Jo, D. (2017). *Better the devil you know: An online field experiment on news consumption.* Retrieved from https://bfi.uchicago.edu/sites/default/files/research/Better_the _Devil_You_Know_Online_Field_Experiment_on_News_Consumption-2.pdf

Johnson, N. R. (1988). "Fire in a crowded theater: A descriptive investigation of the emergence of panic." *International Journal of Mass Emergencies and Disasters, 6*(1), 7–26.

Johnson, N. R., Feinberg, W. E., & Johnston, D. M. (1994). "Microstructure and panic: The impact of social bonds on individual action in collective flight from the Beverly Hills Supper Club fire." In R. R. Dynes & K. J. Tierney (Eds.), *Disasters, collective behavior and social organizations* (pp. 168–189). Newark: University of Delaware Press.

Jordan, S., Brimbal, L., Wallace, D. B., Kassin, S. M., Hartwig, M., & Street, C. N. (In press). "A test of the micro-expressions training tool: Does it improve lie detection?" *Journal of Investigative Psychology and Offender Profiling.* https://doi.org/doi.org/10.1002 /jip.1532

Juergensmeyer, M. (1980). "What if the Untouchables don't believe in Untouchability?" *Bulletin of Concerned Asian Scholars, 12*(1), 23–28.

Kahneman, D. (2011). *Thinking, fast and slow.* New York: Farrar, Straus and Giroux.

Kalla, J. L., & Broockman, D. E. (2018). "The minimal persuasive effects of campaign contact in general elections: Evidence from 49 field experiments." *American Political Science Review, 112*(1), 148–166.

Kallis, A. (2008). *Nazi propaganda in the Second World War.* London: Palgrave Macmillan.

Kam, C. D., & Zechmeister, E. J. (2013). "Name recognition and candidate support." *American Journal of Political Science, 57*(4), 971–986.

Kanwisher, N. (2000). "Domain specificity in face perception." *Nature Neuroscience, 3*(8), 759–763.

Kaplan, S. L. (1982). *Le complot de famine: Histoire d'une rumeur au XVIIIe siècle* (Vol. 39). Paris: A. Colin.

Karsh, E., & Rautsi, I. (2007). *Saddam Hussein: A political biography.* New York: Grove/ Atlantic.

Kassin, S. M., & Gudjonsson, G. H. (2004). "The psychology of confessions: A review of the literature and issues." *Psychological Science in the Public Interest, 5*(2), 33–67.

Kassin, S. M., Meissner, C. A., & Norwick, R. J. (2005). "'I'd know a false confession if I saw one': A comparative study of college students and police investigators." *Law and Human Behavior, 29*(2), 211–227.

Kassin, S. M., & Neumann, K. (1997). "On the power of confession evidence: An experimental test of the fundamental difference hypothesis." *Law and Human Behavior, 21*(5), 469–484.

Kassin, S. M., & Wrightsman, L. S. (1980). "Prior confessions and mock juror verdicts." *Journal of Applied Social Psychology, 10*(2), 133–146.

Katz, E. (1957). "The two-step flow of communication: An up-to-date report on an hypothesis." *Public Opinion Quarterly, 21*(1), 61–78.

Katz, E., & Lazarsfeld, P. F. (1955). *Personal influence: The part played by people in the flow of mass communications.* Glencoe: Free Press.

Kay, J. (2011). *Among the Truthers: A journey through America's growing conspiracist underground.* New York: HarperCollins.

Keil, F. C., Stein, C., Webb, L., Billings, V. D., & Rozenblit, L. (2008). "Discerning the division of cognitive labor: An emerging understanding of how knowledge is clustered in other minds." *Cognitive Science, 32*(2), 259–300.

Keller, K. L., & Lehmann, D. R. (2006). "Brands and branding: Research findings and future priorities." *Marketing Science, 25*(6), 740–759.

Kennedy, J. A., Anderson, C., & Moore, D. A. (2013). "When overconfidence is revealed to others: Testing the status-enhancement theory of overconfidence." *Organizational Behavior and Human Decision Processes, 122*(2), 266–279.

Kennedy, J. J. (2009). "Maintaining popular support for the Chinese Communist Party: The influence of education and the state-controlled media." *Political Studies, 57*(3), 517–536.

Kershaw, I. (1983a). "How effective was Nazi propaganda?" In D. Welch (Ed.), *Nazi propaganda: The power and the limitations* (pp. 180–205). London: Croom Helm.

Kershaw, I. (1983b). *Popular opinion and political dissent in the Third Reich, Bavaria 1933–1945.* New York: Oxford University Press.

Kershaw, I. (1987). *The "Hitler myth": Image and reality in the Third Reich.* New York: Oxford University Press.

Kershaw, I. (1991). *Hitler: Profiles in power.* London: Routledge.

Khare, R. S. (1984). *The untouchable as himself: Ideology, identity and pragmatism among the Lucknow Chamars* (Vol. 8). Cambridge: Cambridge University Press.

Kierkegaard, S. (1961). *Diary* (P. P. Rohde, Ed.). London: Peter Owen.

Kim, G., & Kwak, K. (2011). "Uncertainty matters: Impact of stimulus ambiguity on infant social referencing." *Infant and Child Development, 20*(5), 449–463.

Kim, J. W. (2018). *Evidence can change partisan minds: Rethinking the bounds of motivated reasoning.* Working paper.

Kim, J. W., & Kim, E. (in press). "Identifying the effect of political rumor diffusion using variations in survey timing." *Quarterly Journal of Political Science.*

King, G., Pan, J., & Roberts, M. E. (2017). "How the Chinese government fabricates social media posts for strategic distraction, not engaged argument." *American Political Science Review, 111*(3), 484–501.

Kitcher, P. (1993). *The advancement of science: Science without legend, objectivity without illusions.* New York: Oxford University Press.

Klapper, J. T. (1960). *The effects of mass communication.* Glencoe, IL: Free Press.

Klar, S., Krupnikov, Y., & Ryan, J. B. (2018). "Affective polarization or partisan disdain? Untangling a dislike for the opposing party from a dislike of partisanship." *Public Opinion Quarterly, 82*(2), 379–390.

Klarman, M. J. (2016). *The framers' coup: The making of the United States Constitution.* New York: Oxford University Press.

Knapp, R. H. (1944). "A psychology of rumor." *Public Opinion Quarterly, 8*(1), 22–37.

Knittel, C. R., & Stango, V. (2009). *Shareholder value destruction following the Tiger Woods scandal.* University of California. Retrieved from Faculty. Gsm. Ucdavis. Edu/~ Vstango/Tiger004. Pdf Koch

Kramer, A. D., Guillory, J. E., & Hancock, J. T. (2014). "Experimental evidence of massive-scale emotional contagion through social networks." *Proceedings of the National Academy of Sciences,* 201320040.

Krebs, J. R., & Dawkins, R. (1984). "Animal signals: Mind-reading and manipulation?" In J. Krebs, R., & Davies, N. B. (Eds.), *Behavioural ecology: An evolutionary approach* (Vol. 2, pp. 390–402). Oxford: Basil Blackwell Scientific Publications.

Kuller, C. (2015). "The demonstrations in support of the Protestant provincial bishop Hans Meiser: A successful protest against the Nazi regime." In N. Stoltzfus & B. Maier-Katkin (Eds.), *Protest in Hitler's "National Community": Popular unrest and the Nazi response* (pp. 38–54). New York: Berghahn.

Kurzban, R., & Christner, J. (2011). "Are supernatural beliefs commitment devices for intergroup conflict?" In J. P. Forgas, A. Kruglanski, & K. D. Williwas (Eds.), *The psychology of social conflict and aggression* (pp. 285–300). Sydney Symposium of Social Psychology, vol. 13). New York: Taylor and Francis.

Kushnir, T., Vredenburgh, C., & Schneider, L. A. (2013). "'Who can help me fix this toy?': The distinction between causal knowledge and word knowledge guides preschoolers' selective requests for information." *Developmental Psychology, 49*(3), 446–453.

Lacan, J. (1939). "De l'impulsion au complexe." *Revue Française de Psychanalyse, 1,* 137–141.

Lacan, J. (1970). *Of structure as an inmixing of an otherness prerequisite to any subject whatever* (R. Macksey & E. Donato, Eds.). Baltimore: Johns Hopkins University Press.

Lacan, J. (1980). *De la Psychose paranoïaque dans ses rapports avec la personnalité.* Paris: Seuil.

Lacan, J. (2005). *Le Séminaire, Livre 23, le sinthome.* Paris: Seuil.

Ladd, J. M. (2011). *Why Americans hate the media and how it matters.* New York: Princeton University Press.

Ladd, J. M., & Lenz, G. S. (2009). "Exploiting a rare communication shift to document the persuasive power of the news media." *American Journal of Political Science,* 53(2), 394–410.

Lagrange, P. (2005). *La guerre des mondes at-elle eu lieu?* Paris: Robert Laffont.

Laland, K. N. (2004). "Social learning strategies." *Animal Learning and Behavior,* 32(1), 4–14.

Lanternari, V. (1963). *The religions of the oppressed: A study of modern messianic cults.* New York: Knopf.

Lanzetta, J. T., & Englis, B. G. (1989). "Expectations of cooperation and competition and their effects on observers' vicarious emotional responses." *Journal of Personality and Social Psychology,* 56(4), 543–554.

Larrick, R. P., & Soll, J. B. (2006). "Intuitions about combining opinions: Misappreciation of the averaging principle." *Management Science,* 52, 111–127.

Larson, H. J. (2018). "The biggest pandemic risk? Viral misinformation." *Nature,* 562(7727), 309–309.

Lasswell, H. D. (1927). *Propaganda technique in the world war.* Cambridge, MA: MIT Press.

Laustsen, L., & Bor, A. (2017). "The relative weight of character traits in political candidate evaluations: Warmth is more important than competence, leadership and integrity." *Electoral Studies,* 49, 96–107.

Lawlor, H. J., & Oulton, J. E. L. (1928). *The ecclesiastical history and the martyrs of Palestine: Introduction, notes and index* (Vol. 2). London: Society for Promoting Christian Knowledge.

Lazarsfeld, P. F., Berelson, B., & Gaudet, H. (1948). *The people's choice: How the voter makes up his mind in a presidential campaign.* New York: Columbia University Press.

Le Bon, G. (1897). *The crowd: A study of the popular mind.* London: Macmillian.

Le Bon, G. (1900). *Psychologie des foules.* Paris: Alcan.

Le Bras, G. (1955). *Etudes de sociologie religieuse.* Paris: Presses Universitaires de France.

Leeper, T. J., & Slothuus, R. (2015). *Can citizens be framed? How information, not emphasis, changes opinions.* Unpublished manuscript, Aarhus University.

Leese, D. (2011). *Mao cult: Rhetoric and ritual in China's Cultural Revolution*. Cambridge: Cambridge University Press.

Lenz, G. S. (2009). "Learning and opinion change, not priming: Reconsidering the priming hypothesis." *American Journal of Political Science*, 53(4), 821–837.

Lenz, G. S. (2013). *Follow the leader? How voters respond to politicians' policies and performance*. Chicago: University of Chicago Press.

Le Roy Ladurie, E. (2016). *Montaillou, village occitan de 1294 à 1324*. Paris: Editions Gallimard.

Levendusky, M. S., & Malhotra, N. (2015). "(Mis)perceptions of partisan polarization in the American public." *Public Opinion Quarterly*, 80(S1), 378–391.

Levine, T. R. (2014). "Truth-default theory (TDT): A theory of human deception and deception detection." *Journal of Language and Social Psychology*, 33(4), 378–392.

Lévi-Strauss, C. (1967). "The sorcerer and his magic." In J. Middleton (Ed.), *Magic, witchcraft, and curing* (pp. 23–42). New York: Natural History Press.

Lévi-Strauss, C. (1986). "Entretien avec Judith Miller et Alain Grosrichard." *L'Ane. Le Magazine Freudien*, 20, 27–29.

Lewis, R. A., & Rao, J. M. (2013). *On the near impossibility of measuring the returns to advertising*. Unpublished paper, Google, Inc. and Microsoft Research. Retrieved from http://justinmrao.com/lewis_rao_nearimpossibility.pdf

Liberman, V., Minson, J. A., Bryan, C. J., & Ross, L. (2012). "Naïve realism and capturing the 'wisdom of dyads.'" *Journal of Experimental Social Psychology*, 48(2), 507–512.

Linton, R. (1963). *Acculturation in seven American Indian tribes*. New York: Peter Smith.

Lloyd, G., & Sivin, N. (2002). *The way and the word: Science and medicine in early China and Greece*. New Haven, CT: Yale University Press.

Lodish, L. M., Abraham, M., Kalmenson, S., Livelsberger, J., Lubetkin, B., Richardson, B., & Stevens, M. E. (1995). "How TV advertising works: A meta-analysis of 389 real world split cable TV advertising experiments." *Journal of Marketing Research*, 32(2), 125–139.

Lopez-Ibor, J. J., Soria, J., Canas, F., & Rodriguez-Gamazo, M. (1985). "Psychopathological aspects of the toxic oil syndrome catastrophe." *British Journal of Psychiatry*, 147(4), 352–365.

Luke, T. J. (in press). "Lessons from Pinocchio: Cues to deception may be highly exaggerated." *Perspectives on Psychological Science*, 1745691619838258. https://doi.org/10.1177/1745691619838258

Lull, R. B., & Bushman, B. J. (2015). "Do sex and violence sell? A meta-analytic review of the effects of sexual and violent media and ad content on memory, attitudes, and buying intentions." *Psychological Bulletin*, 141(5), 1022–1048.

Lutz, D. J., & Keil, F. C. (2002). "Early understanding of the division of cognitive labor." *Child Development*, 73(4) 1073–1084.

Macfarlane, A. (1970). "Witchcraft in Tudor and Stuart Essex." In M. Douglas (Ed.), *Witchcraft confessions and accusations* (pp. 81–101). London: Routledge.

MacMullen, R. (1984). *Christianizing the Roman Empire (AD 100–400)*. New Haven, CT: Yale University Press.

MacMullen, R. (1999). *Christianity and paganism in the fourth to eighth centuries*. New Haven, CT: Yale University Press.

Madden, J. R. (2002). "Bower decorations attract females but provoke other male spotted bowerbirds: Bower owners resolve this trade-off." *Proceedings of the Royal Society of London. Series B: Biological Sciences*, 269(1498), 1347–1351.

Maines, L. A. (1990). "The effect of forecast redundancy on judgments of a consensus forecast's expected accuracy." *Journal of Accounting Research*, 28, 29–47.

Majima, Y. (2015). "Belief in pseudoscience, cognitive style and science literacy." *Applied Cognitive Psychology*, 29(4), 552–559.

Malkiel, B. G., & McCue, K. (1985). *A random walk down Wall Street*. New York: Norton.

Malle, B. F., Knobe, J. M., & Nelson, S. E. (2007). "Actor-observer asymmetries in explanations of behavior: New answers to an old question." *Journal of Personality and Social Psychology*, 93(4), 491–514.

Mancosu, P. (1999). "Between Vienna and Berlin: The immediate reception of Godel's incompleteness theorems." *History and Philosophy of Logic*, 20(1), 33–45.

Mansbridge, J. (1999). "Everyday talk in the deliberative system." In S. Macedo (Ed.), *Deliberative politics: Essays on democracy and disagreement* (pp. 211–42). New York: Oxford University Press.

Marcuse, H. (1966). *Eros and civilization: Philosophical inquiry into Freud*. Boston: Beacon Press.

Márquez, X. (2016). *Non-democratic politics: Authoritarianism, dictatorship and democratization*. London: Macmillan International Higher Education.

Márquez, X. (2018). "Two models of political leader cults: Propaganda and ritual." *Politics, Religion and Ideology*, 19(3), 1–20.

Martin, G. J., & Yurukoglu, A. (2017). "Bias in cable news: Persuasion and polarization." *American Economic Review*, 107(9), 2565–2599.

Marx, K., & Engels, F. (1970). *The German ideology*. New York: International Publishers.

Mascaro, O., & Morin, O. (2014). "Gullible's travel: How honest and trustful children become vigilant communicators." In L. Robinson & S. Einav (Eds.), *Trust and skepticism: Children's selective learning from testimony*. London: Psychology Press.

Matsui, T., Rakoczy, H., Miura, Y., & Tomasello, M. (2009). "Understanding of speaker certainty and false-belief reasoning: A comparison of Japanese and German preschoolers." *Developmental Science*, 12(4), 602–613.

Matz, S. C., Kosinski, M., Nave, G., & Stillwell, D. J. (2017). "Psychological targeting as an effective approach to digital mass persuasion." *Proceedings of the National Academy of Sciences*, 114(48), 12714–12719.

Maurer, D. (1999). *The big con: The story of the confidence man*. New York: Anchor Books.

Mawson, A. R. (2012). *Mass panic and social attachment: The dynamics of human behavior*. Aldershot: Ashgate.

Maynard Smith, J., & Harper, D. (2003). *Animal signals*. Oxford: Oxford University Press.

McCloskey, M., Caramazza, A., & Green, B. (1980). "Curvilinear motion in the absence of external forces: Naive beliefs about the motion of objects." *Science*, 210(4474), 1139–1141.

McCloskey, M., Washburn, A., & Felch, L. (1983). "Intuitive physics: The straight-down belief and its origin." *Journal of Experimental Psychology: Learning, Memory, and Cognition*, 9(4), 636–649.

McCullough, M. E., Swartwout, P., Shaver, J. H., Carter, E. C., & Sosis, R. (2016). "Christian religious badges instill trust in Christian and non-Christian perceivers." *Psychology of Religion and Spirituality*, 8(2), 149–163.

McIntyre, L. (2018). *Post-truth*. Cambridge, MA: MIT Press.

McPhail, C. (1991). *The myth of the madding crowd*. New York: Aldine de Gruyter.

McPhail, C. (2007). *A sociological primer on crowd behavior*. Retrieved from https://www.academia.edu/1292597/_2007_A_Sociological_Primer_on_Crowd_Behavior_

McPhail, C., & Wohlstein, R. T. (1983). "Individual and collective behaviors within gatherings, demonstrations, and riots." *Annual Review of Sociology*, 9(1) 579–600.

Meissner, C. A., & Kassin, S. M. (2002). "'He's guilty!': Investigator bias in judgments of truth and deception." *Law and Human Behavior*, 26(5), 469–480.

Meissner, C. A., Surmon-Böhr, F., Oleszkiewicz, S., & Alison, L. J. (2017). "Developing an evidence-based perspective on interrogation: A review of the US government's high-value detainee interrogation group research program." *Psychology, Public Policy, and Law*, 23(4), 438–457.

Mencher, J. P. (1974). "The caste system upside down, or the not-so-mysterious East." *Current Anthropology*, 15(4), 469–493.

Mercier, H. (2011). "Self-deception: Adaptation or by-product?" *Behavioral and Brain Sciences*, 34(1), 35.

Mercier, H. (2012). "Looking for arguments." *Argumentation*, 26(3), 305–324.

Mercier, H. (2013). "Our pigheaded core: How we became smarter to be influenced by other people." In B. Calcott, R. Joyce, & K. Sterelny (Eds.), *Cooperation and its evolution* (pp. 373–398). Cambridge, MA: MIT Press.

Mercier, H. (2016a). "The argumentative theory: Predictions and empirical evidence." *Trends in Cognitive Sciences, 20*(9), 689–700.

Mercier, H. (2016b). "Confirmation (or myside) bias." In R. Pohl (Ed.), *Cognitive Illusions* (2nd ed., pp. 99–114). London: Psychology Press.

Mercier, H. (2017). "How gullible are we? A review of the evidence from psychology and social science." *Review of General Psychology, 21*(2), 103–122.

Mercier, H. (submitted). *The cultural evolution of oaths, ordeals, and lie detectors.*

Mercier, H., Bernard, S., & Clément, F. (2014). "Early sensitivity to arguments: How preschoolers weight circular arguments." *Journal of Experimental Child Psychology, 125*, 102–109.

Mercier, H., Bonnier, P., & Trouche, E. (2016c). "Why don't people produce better arguments?" In L. Macchi, M. Bagassi, & R. Viale (Eds.), *Cognitive Unconscious and Human Rationality* (pp. 205–218). Cambridge, MA: MIT Press.

Mercier, H., Dockendorff, M., & Schwartzberg, M. (submitted). *Democratic legitimacy and attitudes about information-aggregation procedures.*

Mercier, H., Majima, Y., Claidière, N., & Léone, J. (submitted). *Obstacles to the spread of unintuitive beliefs.*

Mercier, H., & Miton, H. (2019). "Utilizing simple cues to informational dependency." *Evolution and Human Behavior, 40*(3), 301–314.

Mercier, H., & Morin, O. (2019). "Majority rules: How good are we at aggregating convergent opinions?" *Evolutionary Human Sciences, 1*, e6.

Mercier, H., & Sperber, D. (2011). "Why do humans reason? Arguments for an argumentative theory." *Behavioral and Brain Sciences, 34*(2), 57–74.

Mercier, H., & Sperber, D. (2017). *The enigma of reason.* Cambridge, MA: Harvard University Press.

Mercier, H., Sudo, M., Castelain, T., Bernard, S., & Matsui, T. (2018). "Japanese preschoolers' evaluation of circular and non-circular arguments." *European Journal of Developmental Psychology, 15*(5), 493–505.

Miguel, E. (2005). "Poverty and witch killing." *Review of Economic Studies, 72*(4), 1153–1172.

Milgram, S. (1974). *Obedience to authority: An experimental view.* New York: Harper and Row.

Milgram, S., Bickman, L., & Berkowitz, L. (1969). "Note on the drawing power of crowds of different size." *Journal of Personality and Social Psychology, 13*(2), 79–82.

Mills, C. M., & Keil, F. C. (2005). "The development of cynicism." *Psychological Science, 16*(5), 385–390.

Mills, C. M., & Keil, F. C. (2008). "Children's developing notions of (im)partiality." *Cognition, 107*(2), 528–551.
Milner, J.-C. (1995). *L'Œuvre claire: Lacan, la science, la philosophie*. Paris: Seuil.
Minson, J. A., Liberman, V., & Ross, L. (2011). "Two to tango." *Personality and Social Psychology Bulletin, 37*(10), 1325–1338.
Mitnick, K. D., & Simon, W. L. (2002). *The art of deception: Controlling the human element of security*. Indianapolis: Wiley.
Miton, H., Claidière, N., & Mercier, H. (2015). "Universal cognitive mechanisms explain the cultural success of bloodletting." *Evolution and Human Behavior, 36*(4), 303–312.
Miton, H., & Mercier, H. (2015). "Cognitive obstacles to pro-vaccination beliefs." *Trends in Cognitive Sciences, 19*(11), 633–636.
Moon, J. W., Krems, J. A., & Cohen, A. B. (2018). "Religious people are trusted because they are viewed as slow life-history strategists." *Psychological Science*, 0956797617753606.
Moorehead, A. (1965). *African trilogy: The North African campaign 1940–43*. London: Hamish Hamilton.
Morgan, A. C., Economou, D., Way, S. F., & Clauset, A. (2018). *Prestige drives epistemic inequality in the diffusion of scientific ideas*. Retrieved from https://arxiv.org/abs/1805.09966
Morgan, T.J.H., Laland, K. N., & Harris, P. L. (2015). "The development of adaptive conformity in young children: Effects of uncertainty and consensus." *Developmental Science, 18*(4), 511–524.
Morgan, T.J.H., Rendell, L. E., Ehn, M., Hoppitt, W., & Laland, K. N. (2012). "The evolutionary basis of human social learning." *Proceedings of the Royal Society of London B: Biological Sciences, 279*(1729), 653–662.
Morin, E. (1969). *La Rumeur d'Orléans*. Paris: Seuil.
Morin, O. (2016). *How traditions live and die*. New York: Oxford University Press.
Morton-Williams, P. (1956). "The Atinga cult among the south-western Yoruba: A sociological analysis of a witch-finding movement." *Bulletin de l'Institut Français d'Afrique Noire, Série B Sciences Humaines, 18*, 315–334.
Moscovici, S. (1981). *L'Age des foules*. Paris: Fayard.
Moscovici, S. (1985). "Social influence and conformity." In G. Lindzey & E. Aronson (Eds.), *Handbook of social psychology* (3rd ed., Vol. 2, pp. 347–412). New York: Random House.
Murdock, G. P., Wilson, S. F., & Frederick, V. (1978). "World distribution of theories of illness." *Ethnology, 17*, 449–470.
Murray, A. (1974). "Religion among the poor in thirteenth-century France: The testimony of Humbert de Romans." *Traditio, 30*, 285–324.

Myers, D. G. (2009). *Social psychology* (10th ed.). New York: McGraw-Hill.

Myers, D. G., & Bach, P. J. (1974). "Discussion effects on militarism-pacifism: A test of the group polarization hypothesis." *Journal of Personality and Social Psychology, 30*(6), 741–747.

Nadeau, R., Nevitte, N., Gidengil, E., & Blais, A. (2008). "Election campaigns as information campaigns: Who learns what and does it matter?" *Political Communication, 25*(3), 229–248.

Nair, G. (2018). "Misperceptions of relative affluence and support for international redistribution." *Journal of Politics, 80*(3), 815–830.

Naughton, T. J. (1996). "Relationship of personal and situational factors to managers' expectations of organizational change." *Psychological Reports, 78*(1), 313–314.

Nelissen, R. M., & Meijers, M. H. (2011). "Social benefits of luxury brands as costly signals of wealth and status." *Evolution and Human Behavior, 32*(5), 343–355.

Nichols, S. (2002). "On the genealogy of norms: A case for the role of emotion in cultural evolution." *Philosophy of Science, 69*(2), 234–255.

Nishida, N., Yano, H., Nishida, T., Kamura, T., & Kojiro, M. (2006). "Angiogenesis in cancer." *Vascular Health and Risk Management, 2*(3), 213–219.

Nitecki, M. H., Lemke, J. L., Pullman, H. W., & Johnson, M. E. (1978). "Acceptance of plate tectonic theory by geologists." *Geology, 6*(11), 661–664.

Norscia, I., & Palagi, E. (2011). "Yawn contagion and empathy in Homo sapiens." *PloS One, 6*(12), e28472.

Nunn, N., & Sanchez de la Sierra, R. (2017). "Why being wrong can be right: Magical warfare technologies and the persistence of false beliefs." *American Economic Review, 107*(5), 582–587.

Nyhan, B., Porter, E., Reifler, J., & Wood, T. (2017). *Taking corrections literally but not seriously? The effects of information on factual beliefs and candidate favorability.* Unpublished manuscript.

Nyhan, B., & Reifler, J. (2010). "When corrections fail: The persistence of political misperceptions." *Political Behavior, 32*(2), 303–330.

Nyhan, B., & Reifler, J. (2015). "Does correcting myths about the flu vaccine work? An experimental evaluation of the effects of corrective information." *Vaccine, 33*(3), 459–464.

O'Donnell, V., & Jowett, G. S. (1992). *Propaganda and persuasion.* New York: Sage.

Ong, A. (1987). *Spirits of resistance and capitalist discipline, Second Edition: Factory women in Malaysia.* Albany: SUNY Press.

Open Science Collaboration. (2015). "Estimating the reproducibility of psychological science." *Science, 349*(6251), aac4716.

Oreskes, N. (1988). "The rejection of continental drift." *Historical Studies in the Physical and Biological Sciences, 18*(2), 311–348.

Origgi, G. (2017). *Reputation: What it is and why it matters*. Princeton, NJ: Princeton University Press.

Osnos, E. (2014). *Age of ambition: Chasing fortune, truth, and faith in the new China*. London: Macmillan.

Ostreiher, R., & Heifetz, A. (2017). "The sentinel behaviour of Arabian babbler floaters." *Royal Society Open Science*, 4(2), 160738.

Ostrom, E., Walker, J., & Gardner, R. (1992). "Covenants with and without a sword: Self-governance is possible." *American Political Science Review*, 86(2), 404–417.

Owren, M. J., & Bachorowski, J.-A. (2001). "The evolution of emotional experience: A 'selfish-gene' account of smiling and laughter in early hominids and humans." In T. J. Mayne & G. A. Bonanno (Eds.), *Emotions: Current issues and future directions* (pp. 152–191). New York: Guilford Press.

Parker, K., & Jaudel, E. (1989). *Police cell detention in Japan: The Daiyo Kangoku system: A report*. San Francisco: Association of Humanitarian Lawyers.

Peires, J. B. (1989). *The dead will arise: Nongqawuse and the great Xhosa cattle-killing movement of 1856–7*. Bloomington: Indiana University Press.

Peisakhin, L., & Rozenas, A. (2018). "Electoral effects of biased media: Russian television in Ukraine." *American Journal of Political Science*, 62(3), 535–550.

Pennycook, G., Cheyne, J. A., Barr, N., Koehler, D. J., & Fugelsang, J. A. (2015). "On the reception and detection of pseudo-profound bullshit." *Judgment and Decision Making*, 10(6), 549–563.

Pennycook, G., Cheyne, J. A., Seli, P., Koehler, D. J., & Fugelsang, J. A. (2012). "Analytic cognitive style predicts religious and paranormal belief." *Cognition*, 123(3), 335–346.

Pennycook, G., & Rand, D. G. (2018). "Lazy, not biased: Susceptibility to partisan fake news is better explained by lack of reasoning than by motivated reasoning." *Cognition*, 188, 39–50.

Perry, G. (2013). *Behind the shock machine: The untold story of the notorious Milgram psychology experiments*. New York: New Press.

Perry, G., Brannigan, A., Wanner, R. A., & Stam, H. (In press). "Credibility and incredulity in Milgram's obedience experiments: A reanalysis of an unpublished test." *Social Psychology Quarterly*. https://doi.org/10.1177/0190272519861952

Petersen, M. B., Osmundsen, M., & Arceneaux, K. (2018). *A "need for chaos" and the sharing of hostile political rumors in advanced democracies*. https://doi.org/10.31234/osf.io/6m4ts

Peterson, J. B. (2002). *Maps of meaning: The architecture of belief*. London: Routledge.

Petrocelli, J. V. (2018). "Antecedents of bullshitting." *Journal of Experimental Social Psychology*, 76, 249–258.

Petrova, M., & Yanagizawa-Drott, D. (2016). "Media persuasion, ethnic hatred, and mass violence." In C. H. Anderton & J. Brauer (Eds.), *Economic aspects of genocides, other mass atrocities, and their prevention* (p. 274–286). Oxford: Oxford University Press.

Pettegree, A. (2014). *The invention of news: How the world came to know about itself*. New Haven, CT: Yale University Press.

Petty, R. E., & Wegener, D. T. (1998). "Attitude change: Multiple roles for persuasion variables." In D. T. Gilbert, S. Fiske, & G. Lindzey (Eds.), *The handbook of social psychology* (pp. 323–390). Boston: McGraw-Hill.

Pfaff, S. (2001). "The limits of coercive surveillance: Social and penal control in the German Democratic Republic." *Punishment and Society, 3*(3), 381–407.

Pinker, S. (1997). *How the mind works*. New York: Norton.

Planck, M. (1968). *Scientific autobiography and other papers* (F. Gaynor, Trans.). New York: Citadel Press.

Platow, M. J., Foddy, M., Yamagishi, T., Lim, L., & Chow, A. (2012). "Two experimental tests of trust in in-group strangers: The moderating role of common knowledge of group membership." *European Journal of Social Psychology, 42*(1), 30–35.

Pomper, G. M., & Lederman, S. S. (1980). *Elections in America: Control and influence in democratic politics*. New York: Longman.

Porter, S., & ten Brinke, L. (2008). "Reading between the lies: Identifying concealed and falsified emotions in universal facial expressions." *Psychological Science, 19*(5), 508–514.

Pound, J., & Zeckhauser, R. (1990). "Clearly heard on the street: The effect of takeover rumors on stock prices." *Journal of Business, 63*(3), 291–308.

Power, E. A. (2017). "Social support networks and religiosity in rural South India." *Nature Human Behaviour, 1*(3), 0057.

Prasad, J. (1935). "The psychology of rumour: A study relating to the great Indian earthquake of 1934." *British Journal of Psychology. General Section, 26*(1), 1–15.

Pratkanis, A. R., & Aronson, E. (1992). *Age of propaganda: The everyday use and abuse of persuasion*. New York: W. H. Freeman.

Priniski, J., & Horne, Z. (2018). "Attitude change on Reddit's change my view." *Proceedings of the Cognitive Science Society Conference*.

Proulx, G., Fahy, R. F., & Walker, A. (2004). *Analysis of first-person accounts from survivors of the World Trade Center evacuation on September 11*. Retrieved from https://s3.amazonaws.com/academia.edu.documents/36860616/Analysis_of_First-Person_Accounts.PDF?AWSAccessKeyId=AKIAIWOWYYGZ2Y53UL3A&Expires=1542920752&Signature=S5zsNHIA%2BObbcYJA%2BSBpXT%2BGrR8%3D&response-content-disposition=inline%3B%20filename%3DAnalysis_of_First-Person_Accounts_PDF.pdf

Pulford, B. D., Colman, A. M., Buabang, E. K., & Krockow, E. M. (2018). "The persuasive power of knowledge: Testing the confidence heuristic." *Journal of Experimental Psychology: General, 147*(10), 1431–1444.

Puschmann, C. (2018, November). "Beyond the bubble: Assessing the diversity of political search results." *Digital Journalism*, doi: https://doi.org/10.1080/21670811.2018.1539626

Radelet, M. L., Bedau, H. A., & Putnam, C. E. (1994). *In spite of innocence: Erroneous convictions in capital cases*. Boston: Northeastern University Press.

Rankin, P. J., & Philip, P. J. (1963). "An epidemic of laughing in the Bukoba district of Tanganyika." *Central African Journal of Medicine*, 9(5), 167–170.

Raskin, D. C., Honts, C. R., & Kircher, J. C. (2013). *Credibility assessment: Scientific research and applications*. London: Academic Press.

Ratcliffe, J. M., Fenton, M. B., & Galef, B. G., Jr. (2003). "An exception to the rule: Common vampire bats do not learn taste aversions." *Animal Behaviour*, 65(2), 385–389.

Reed, L. I., DeScioli, P., & Pinker, S. A. (2014). "The commitment function of angry facial expressions." *Psychological Science*, 25(8), 1511–1517.

Reicher, S. D. (1996). "'The Crowd' century: Reconciling practical success with theoretical failure." *British Journal of Social Psychology*, 35(4), 535–553.

Reicher, S. D., Haslam, S. A., & Smith, J. R. (2012). "Working toward the experimenter: Reconceptualizing obedience within the Milgram paradigm as identification-based followership." *Perspectives on Psychological Science*, 7(4), 315–324.

Reid, T. (1970). *Inquiry into the human mind*. Chicago: University of Chicago Press. (Original work published 1764.)

Reyes-Jaquez, B., & Echols, C. H. (2015). "Playing by the rules: Self-interest information influences children's trust and trustworthiness in the absence of feedback." *Cognition*, 134, 140–154.

Richerson, P. J., & Boyd, R. (2005). *Not by genes alone*. Chicago: University of Chicago Press.

Richter, T., Schroeder, S., & Wöhrmann, B. (2009). "You don't have to believe everything you read: Background knowledge permits fast and efficient validation of information." *Journal of Personality and Social Psychology*, 96(3), 538–558.

Robbins, T. (1988). *Cults, converts and charisma: The sociology of new religious movements*. New York: Sage.

Roberts, M. E. (2018). *Censored: Distraction and diversion inside China's great firewall*. Princeton, NJ: Princeton University Press.

Robertson, R. E., Jiang, S., Joseph, K., Friedland, L., Lazer, D., & Wilson, C. (2018). "Auditing partisan audience bias within Google search." *Proceedings of the ACM on Human-Computer Interaction*, 2(CSCW). Retrieved from https://dl.acm.org/citation.cfm?id=3274417

Robertson, T. E., Sznycer, D., Delton, A. W., Tooby, J., & Cosmides, L. (2018). "The true trigger of shame: Social devaluation is sufficient, wrongdoing is unnecessary." *Evolution and Human Behavior*, 39(5), 566–573.

Robinson, E. J., Champion, H., & Mitchell, P. (1999). "Children's ability to infer utterance veracity from speaker informedness." *Developmental Psychology, 35*(2), 535–546.

Robinson, F. G. (1988). "The characterization of Jim in *Huckleberry Finn*." *Nineteenth-Century Literature, 43*(3), 361–391.

Robisheaux, T. W. (2009). *The last witch of Langenburg: Murder in a German village*. New York: Norton.

Rocher, L. (1964). "The theory of proof in ancient Hindu law." *Recueil de La Société Jean Bodin, 18*, 325–371.

Rogers, T., & Nickerson, D. (2013). *Can inaccurate beliefs about incumbents be changed? And can reframing change votes?* Retrieved from https://papers.ssrn.com/sol3/papers.cfm?abstract_id=2271654

Rose, R., Mishler, W. T., & Munro, N. (2011). *Popular support for an undemocratic regime: The changing views of Russians*. https://doi.org/10.1017/CBO9780511809200

Rosnow, R. L. (1991). "Inside rumor: A personal journey." *American Psychologist, 46*(5), 484–496.

Rothbard, M. N. (2003). *The ethics of liberty*. New York: NYU Press.

Roulin, N., & Ternes, M. (2019). "Is it time to kill the detection wizard? Emotional intelligence does not facilitate deception detection." *Personality and Individual Differences, 137*, 131–138.

Rousseau, J.-J. (2002). *The social contract: And, the first and second discourses* (G. May, Trans.). New Haven, CT: Yale University Press.

Roy, O. (2016). *Le djihad et la mort*. Paris: Le Seuil.

Royed, T. J. (1996). "Testing the mandate model in Britain and the United States: Evidence from the Reagan and Thatcher eras." *British Journal of Political Science, 26*(1), 45–80.

Rozin, P. (1976). "The selection of foods by rats, humans, and other animals." In R. A. Rosenblatt, A. Hind, E. Shaw, & C. Beer (Eds.), *Advances in the study of behavior* (Vol. 6, pp. 21–76). New York: Academic Press.

Rudé, G. (1959). *The crowd in the French Revolution*. Oxford: Oxford University Press.

Rumsey, A., & Niles, D. (Eds.). (2011). *Sung tales from the Papua New Guinea highlands: Studies in form, meaning, and sociocultural context*. Camberra: ANU E Press.

Sadler, O., & Tesser, A. (1973). "Some effects of salience and time upon interpersonal hostility and attraction during social isolation." *Sociometry, 36*(1), 99–112.

Safra, L., Baumard, N., & Chevallier, C. (submitted). *Why would anyone elect an untrustworthy and narcissistic leader*.

Sala, G., Aksayli, N. D., Tatlidil, K. S., Tatsumi, T., Gondo, Y., & Gobet, F. (2018). "Near and far transfer in cognitive training: A second-order meta-analysis." *Collabra: Psychology, 5*, 18. DOI: doi:10.1525/collabra.203

Sala, G., & Gobet, F. (2017). "Does far transfer exist? Negative evidence from chess, music, and working memory training." *Current Directions in Psychological Science, 26*(6), 515–520.

Sala, G., & Gobet, F. (2018). "Cognitive training does not enhance general cognition." *Trends in Cognitive Sciences, 23*(1), 9–20.

Sally, D. (1995). "Conversation and cooperation in social dilemmas." *Rationality and Society, 7*(1), 58–92.

Salter, S. (1983). "Structures of consensus and coercion: Workers' morale and the maintenance of work discipline, 1939–1945." In D. Welch (Ed.), *Nazi propaganda: The power and the limitations* (pp. 88–116). London: Croom Helm.

San Roque, L., & Loughnane, R. (2012). "The New Guinea Highlands evidentiality area." *Linguistic Typology, 16*(1), 111–167.

Schieffelin, B. B. (1995). "Creating evidence." *Pragmatics: Quarterly Publication of the International Pragmatics Association, 5*(2), 225–243.

Schniter, E., Sheremeta, R. M., & Sznycer, D. (2013). "Building and rebuilding trust with promises and apologies." *Journal of Economic Behavior and Organization, 94,* 242–256.

Schroeder, E., & Stone, D. F. (2015). Fox News and political knowledge. *Journal of Public Economics, 126,* 52–63.

Schultz, D. P. (1964). *Panic behavior: Discussion and readings* (Vol. 28). New York: Random House.

Schweingruber, D., & Wohlstein, R. T. (2005). "The madding crowd goes to school: Myths about crowds in introductory sociology textbooks." *Teaching Sociology, 33*(2), 136–153.

Scott, J. C. (1990). *Domination and the arts of resistance: Hidden transcripts.* New Haven, CT: Yale University Press.

Scott, J. C. (2008). *Weapons of the weak: Everyday forms of peasant resistance.* New Haven, CT: Yale University Press.

Scott-Phillips, T. C. (2008). "Defining biological communication." *Journal of Evolutionary Biology, 21*(2), 387–395.

Scott-Phillips, T. C. (2014). *Speaking our minds: Why human communication is different, and how language evolved to make it special.* London: Palgrave Macmillan.

Scott-Phillips, T. C., Blythe, R. A., Gardner, A., & West, S. A. (2012). "How do communication systems emerge?" *Proceedings of the Royal Society B: Biological Sciences, 279*(1735), 1943–1949.

Seabright, P. (2004). *The company of strangers: A natural history of economic life.* Princeton: Princeton University Press.

Sebestyen, V. (2009). *Revolution 1989: The fall of the Soviet empire.* London: Hachette UK.

Selb, P., & Munzert, S. (2018). "Examining a most likely case for strong campaign effects: Hitler's speeches and the rise of the Nazi Party, 1927–1933." *American Political Science Review*, 112(4), 1050–1066.

Sell, A., Tooby, J., & Cosmides, L. (2009). "Formidability and the logic of human anger." *Proceedings of the National Academy of Sciences*, 106(35), 15073–15078.

Seyfarth, R. M., Cheney, D. L., & Marler, P. (1980). "Vervet monkey alarm calls: Semantic communication in a free-ranging primate." *Animal Behaviour*, 28(4), 1070–1094.

Shea, N., Boldt, A., Bang, D., Yeung, N., Heyes, C., & Frith, C. D. (2014). "Suprapersonal cognitive control and metacognition." *Trends in Cognitive Sciences*, 18(4), 186–193.

Shibutani, T. (1966). *Improvised news: A sociological study of rumor*. New York: Bobbs-Merrill.

Shore, J., Baek, J., & Dellarocas, C. (2018). "Network structure and patterns of information diversity on Twitter." *MIS Quarterly*, 42(3), 849–872.

Shtulman, A. (2006). "Qualitative differences between naïve and scientific theories of evolution." *Cognitive Psychology*, 52(2), 170–194.

Shtulman, A. (2017). *Scienceblind: Why our intuitive theories about the world are so often wrong*. New York: Basic Books.

Shtulman, A., & Valcarcel, J. (2012). "Scientific knowledge suppresses but does not supplant earlier intuitions." *Cognition*, 124(2), 209–215.

Sighele, S. (1901). *La foule criminelle: Essai de psychologie collective*. Paris: Alcan.

Signer, M. (2009). *Demagogue: The fight to save democracy from its worst enemies*. New York: Macmillan.

Sigurdsson, J. F., & Gudjonsson, G. H. (1996). "The psychological characteristics of 'false confessors': A study among Icelandic prison inmates and juvenile offenders." *Personality and Individual Differences*, 20(3), 321–329.

Silver, B. (1987). "Political beliefs of the Soviet citizen: Sources of support for regime norms." In J. R. Millar (Ed.), *Politics, work, and daily life in the USSR*. New York: Cambridge University Press.

Silver, I., & Shaw, A. (2018). "No harm, still foul: Concerns about reputation drive dislike of harmless plagiarizers." *Cognitive Science*, 42(S1), 213–240.

Simler, K., & Hanson, R. (2017). *The elephant in the brain: Hidden motives in everyday life*. New York: Oxford University Press.

Singh, M. (2018). "The cultural evolution of shamanism." *Behavioral and Brain Sciences*, 41, e66.

Sinha, D. (1952). "Behaviour in a catastrophic situation: A psychological study of reports and rumours." *British Journal of Psychology. General Section*, 43(3), 200–209.

Sklar, A. Y., Levy, N., Goldstein, A., Mandel, R., Maril, A., & Hassin, R. R. (2012). "Reading and doing arithmetic nonconsciously." *Proceedings of the National Academy of Sciences, 109*(48), 19614–19619.

Smith, E. A., & Bird, R.L.B. (2000). "Turtle hunting and tombstone opening: Public generosity as costly signaling." *Evolution and Human Behavior, 21*(4), 245–261.

Smith, M. J., Ellenberg, S. S., Bell, L. M., & Rubin, D. M. (2008). "Media coverage of the measles-mumps-rubella vaccine and autism controversy and its relationship to MMR immunization rates in the United States." *Pediatrics, 121*(4), e836–e843.

Sniezek, J. A., Schrah, G. E., & Dalal, R. S. (2004). "Improving judgement with prepaid expert advice." *Journal of Behavioral Decision Making, 17*(3), 173–190.

Snow, David A., & Phillips, C. L. (1980). "The Lofland-Stark conversion model: A critical reassessment." *Social Problems, 27*(4), 430–447.

Snyder, J. M., & Strömberg, D. (2010). "Press coverage and political accountability." *Journal of Political Economy, 118*(2), 355–408.

Sodian, B., Thoermer, C., & Dietrich, N. (2006). "Two- to four-year-old children's differentiation of knowing and guessing in a non-verbal task." *European Journal of Developmental Psychology, 3*(3), 222–237.

Sokal, A. D., & Bricmont, J. (1998). *Intellectual impostures: Postmodern philosophers' abuse of science*. London: Profile Books.

Sommer, C. (2011). "Alarm calling and sentinel behaviour in Arabian babblers." *Bioacoustics, 20*(3), 357–368.

Sperber, D. (1975). *Rethinking symbolism*. Cambridge: Cambridge University Press.

Sperber, D. (1994). "The modularity of thought and the epidemiology of representations." In L. A. Hirschfeld & S. A. Gelman (Eds.), *Mapping the mind: Domain specificity in cognition and culture* (pp. 39–67). Cambridge: Cambridge University Press.

Sperber, D. (1997). "Intuitive and reflective beliefs." *Mind and Language, 12*(1), 67–83.

Sperber, D. (2010). "The guru effect." *Review of Philosophy and Psychology, 1*(4), 583–592.

Sperber, D., & Baumard, N. (2012). "Moral reputation: An evolutionary and cognitive perspective." *Mind and Language, 27*(5), 495–518.

Sperber, D., Clément, F., Heintz, C., Mascaro, O., Mercier, H., Origgi, G., & Wilson, D. (2010). "Epistemic vigilance." *Mind and Language, 25*(4), 359–393.

Sperber, D., & Mercier, H. (2018). "Why a modular approach to reason?" *Mind and Language, 131*(4), 496–501.

Sperber, D., & Wilson, D. (1995). *Relevance: Communication and cognition*. New York: Wiley-Blackwell.

Stanley, J. (2015). *How propaganda works*. New York: Princeton University Press.

Stapleton, T. J. (1991). "'They no longer care for their chiefs': Another look at the Xhosa cattle-killing of 1856–1857." *International Journal of African Historical Studies, 24*(2), 383–392.

Stark, R. (1984). The rise of a new world faith. *Review of Religious Research, 26*(1), 18–27.

Stark, R. (1996). *The rise of Christianity: A sociologist reconsiders history*. Princeton, NJ: Princeton University Press.

Stark, R. (1999). "Secularization, RIP." *Sociology of Religion, 60*(3), 249–273.

Stark, R., & Bainbridge, W. S. (1980). "Networks of faith: Interpersonal bonds and recruitment to cults and sects." *American Journal of Sociology, 85*(6), 1376–1395.

Stenberg, G. (2013). "Do 12-month-old infants trust a competent adult?" *Infancy, 18*(5), 873–904.

Sterelny, K. (2012). *The evolved apprentice*. Cambridge, MA: MIT Press.

Sternberg, R. J. (1985). *Beyond IQ: A triarchic theory of human intelligence*. Cambridge: Cambridge University Press.

Stibbard-Hawkes, D. N., Attenborough, R. D., & Marlowe, F. W. (2018). "A noisy signal: To what extent are Hadza hunting reputations predictive of actual hunting skills?" *Evolution and Human Behavior, 39*(6), 639–651.

Stimson, J. A. (2004). *Tides of consent: How public opinion shapes American politics*. Cambridge: Cambridge University Press.

Stone, J. R. (2016). *The craft of religious studies*. New York: Springer.

Stout, M. J. (2011). *The effectiveness of Nazi propaganda during World War II* (master's thesis). Eastern Michigan University.

Strahan, E. J., Spencer, S. J., & Zanna, M. P. (2002). "Subliminal priming and persuasion: Striking while the iron is hot." *Journal of Experimental Social Psychology, 38*(6), 556–568.

Strandburg-Peshkin, A., Farine, D. R., Couzin, I. D., & Crofoot, M. C. (2015). "Shared decision-making drives collective movement in wild baboons." *Science, 348*(6241), 1358–1361.

Strauss, C., & Quinn, N. (1997). *A cognitive theory of cultural meaning*. Cambridge: Cambridge University Press.

Street, C. N. H., & Richardson, D. C. (2015). "Lies, damn lies, and expectations: How base rates inform lie-truth judgments." *Applied Cognitive Psychology, 29*(1), 149–155.

Strömberg, D. (2004). Radio's impact on public spending. *Quarterly Journal of Economics, 119*(1), 189–221.

Stroud, N. J., & Lee, J. K. (2013). "Perceptions of cable news credibility." *Mass Communication and Society, 16*(1), 67–88.

Sunstein, C. R. (2018). *#Republic: Divided democracy in the age of social media*. New York: Princeton University Press.

Surowiecki, J. (2005). *The wisdom of crowds*. New York: Anchor Books.
Svolik, M. W. (2012). *The politics of authoritarian rule*. Cambridge: Cambridge University Press.
Sznycer, D., Schniter, E., Tooby, J., & Cosmides, L. (2015). "Regulatory adaptations for delivering information: The case of confession." *Evolution and Human Behavior, 36*(1), 44–51.
Sznycer, D., Xygalatas, D., Agey, E., Alami, S., An, X.-F., Ananyeva, K. I., . . . Flores, C. (2018). "Cross-cultural invariances in the architecture of shame." *Proceedings of the National Academy of Sciences, 115*(39), 9702–9707.
Taber, C. S., & Lodge, M. (2006). "Motivated skepticism in the evaluation of political beliefs." *American Journal of Political Science, 50*(3), 755–769.
Taine, H. (1876). *The origins of contemporary France*. London: H. Holt.
Taine, H. (1885). *The French Revolution* (Vol. 1). London: H. Holt.
Taleb, N. (2005). *Fooled by randomness: The hidden role of chance in life and in the markets*. New York: Random House.
Tamis-LeMonda, C. S., Adolph, K. E., Lobo, S. A., Karasik, L. B., Ishak, S., & Dimitropoulou, K. A. (2008). "When infants take mothers' advice: 18-month-olds integrate perceptual and social information to guide motor action." *Developmental Psychology, 44*(3), 734–746.
Tappin, B. M., & Gadsby, S. (2019). "Biased belief in the Bayesian brain: A deeper look at the evidence." *Consciousness and Cognition, 68*, 107–114.
Tappin, B. M., & McKay, R. T. (2019). "Moral polarization and out-party hostility in the US political context." *Journal of Social and Political Psychology, 7*(1), 213–245.
Tarde, G. (1892). "Les crimes des foules." *Archives de l'Anthropologie Criminelle, 7*, 353–386.
Tarde, G. (1900). *Les lois de l'imitation: Étude sociologique*. Paris: Alcan.
Tellis, G. J. (1988). "Advertising exposure, loyalty, and brand purchase: A two-stage model of choice." *Journal of Marketing Research, 25*(2), 134–144.
Tellis, G. J. (2003). *Effective advertising: Understanding when, how, and why advertising works*. London: Sage.
Tellis, G. J., Chandy, R., & Thaivanich, P. (2000). "Decomposing the effects of direct advertising: Which brand works, when, where, and how long?" *Journal of Marketing Research, 37*(1), 32–46.
ten Brinke, L., MacDonald, S., Porter, S., & O'Connor, B. (2012). "Crocodile tears: Facial, verbal and body language behaviours associated with genuine and fabricated remorse." *Law and Human Behavior, 36*(1), 51–59.
Tenney, E. R., MacCoun, R. J., Spellman, B. A., & Hastie, R. (2007). "Calibration trumps confidence as a basis for witness credibility." *Psychological Science, 18*(1), 46–50.

Tenney, E. R., Small, J. E., Kondrad, R. L., Jaswal, V. K., & Spellman, B. A. (2011). "Accuracy, confidence, and calibration: How young children and adults assess credibility." *Developmental Psychology*, 47(4), 1065.

Tenney, E. R., Spellman, B. A., & MacCoun, R. J. (2008). "The benefits of knowing what you know (and what you don't): How calibration affects credibility." *Journal of Experimental Social Psychology*, 44(5), 1368–1375.

Terrier, N., Bernard, S., Mercier, H., & Clément, F. (2016). "Visual access trumps gender in 3- and 4-year-old children's endorsement of testimony." *Journal of Experimental Child Psychology*, 146, 223–230.

Tesser, A. (1978). "Self-generated attitude change." In L. Berkowitz (Ed.), *Advances in Experimental Social Psychology* (pp. 289–338). New York: Academic Press.

Thagard, P. (2005). "Testimony, credibility, and explanatory coherence." *Erkenntnis*, 63(3), 295–316.

Thomas, K. (1971). *Religion and the decline of magic*. London: Weidenfeld and Nicolson.

Thomas, M. (in press). "Was television responsible for a new generation of smokers?" *Journal of Consumer Research*. https://doi.org/10.1093/jcr/ucz024

Thorndike, E. L. (1917). *The principles of teaching*. New York: AG Seiler.

Tilly, L., & Tilly, R. (1975). *The rebellious century, 1830–1930*. Cambridge: Cambridge University Press.

Tismaneanu, V. (1989). "The tragicomedy of Romanian communism." *East European Politics and Societies*, 3(2), 329–376.

Todorov, A., Funk, F., & Olivola, C. Y. (2015). "Response to Bonnefon et al.: Limited 'kernels of truth' in facial inferences." *Trends in Cognitive Sciences*, 19(8), 422–423.

Tomasello, M., Call, J., & Gluckman, A. (1997). "Comprehension of novel communicative signs by apes and human children." *Child Development*, 68(6), 1067–1080.

Tooby, J., Cosmides, L., & Price, M. E. (2006). "Cognitive adaptations for *n*-person exchange: The evolutionary roots of organizational behavior." *Managerial and Decision Economics*, 27(2–3), 103–129.

Torrey, N. L. (1961). *Les Philosophes: The philosophers of the Enlightenment and modern democracy*. New York: Capricorn Books.

Trappey, C. (1996). "A meta-analysis of consumer choice and subliminal advertising." *Psychology and Marketing*, 13(5), 517–530.

Trouche, E., Sander, E., & Mercier, H. (2014). "Arguments, more than confidence, explain the good performance of reasoning groups." *Journal of Experimental Psychology: General*, 143(5), 1958–1971.

Trouche, E., Shao, J., & Mercier, H. (2019). "How is argument evaluation biased?" *Argumentation*, 33(1), 23–43.

Turner, P. A. (1992). "Ambivalent patrons: The role of rumor and contemporary legends in African-American consumer decisions." *Journal of American Folklore*, 105(418), 424–441.

Turner, R. H. (1964). "Collective behavior." In R.E.L. Paris (Ed.), *Handbook of modern sociology* (pp. 382–425). Chicago: Rand McNally.

Turner, R. H., & Killian, L. M. (1972). *Collective behavior*. Englewood Cliffs, NJ: Prentice-Hall.

Tyndale-Biscoe, C. H. (2005). *Life of marsupials*. Clayton: CSIRO Publishing.

Ullmann, W. (1946). "Medieval principles of evidence." *Law Quarterly Review*, 62, 77–87.

Umbres, R. (2018). *Epistemic vigilance and the social mechanisms of mirthful deception in fool's errands*. Manuscript in preparation.

Underwood, R. H. (1995). "Truth verifiers: From the hot iron to the lie detector." *Kentucky Law Journal*, 84, 597–642.

VanderBorght, M., & Jaswal, V. K. (2009). "Who knows best? Preschoolers sometimes prefer child informants over adult informants." *Infant and Child Development: An International Journal of Research and Practice*, 18(1), 61–71.

van der Linden, S., Maibach, E., & Leiserowitz, A. (2019, May). "Exposure to scientific consensus does not cause psychological reactance." *Environmental Communication*, DOI: https://doi.org/10.1080/17524032.2019.1617763

Van Doorn, G., & Miloyan, B. (2017). "The Pepsi paradox: A review." *Food Quality and Preference*, 65, 194–197.

van Prooijen, J.-W., & Van Vugt, M. (2018). "Conspiracy theories: Evolved functions and psychological mechanisms." *Perspectives on Psychological Science*, 13(6), 770–788.

Van Zant, A. B., & Andrade, E. B. (submitted). "Is there a 'voice' of certainty? Speakers' certainty is detected through paralanguage."

Vargo, C. J., Guo, L., & Amazeen, M. A. (2018). "The agenda-setting power of fake news: A big data analysis of the online media landscape from 2014 to 2016." *New Media and Society*, 20(5), 2028–2049.

Veyne, P. (2002). "Lisibilité des images, propagande et apparat monarchique dans l'Empire romain." *Revue Historique*, 621(1), 3–30.

Vinokur, A. (1971). "Review and theoretical analysis of the effects of group processes upon individual and group decisions involving risk." *Psychological Bulletin*, 76(4), 231–250.

Voigtländer, N., & Voth, H.-J. (2014). *Highway to Hitler*. NBER Working Paper No. 20150. Retrieved from https://www.nber.org/papers/w20150

Voigtländer, N., & Voth, H.-J. (2015). "Nazi indoctrination and anti-Semitic beliefs in Germany." *Proceedings of the National Academy of Sciences*, 112(26), 7931–7936.

von Hippel, W., & Trivers, R. (2011). "The evolution and psychology of self-deception." *Behavioral and Brain Sciences*, 34(1), 1–16.

Vosoughi, S., Roy, D., & Aral, S. (2018). "The spread of true and false news online." *Science*, 359(6380), 1146–1151.

Vrij, A. (2000). *Detecting lies and deceit: The psychology of lying and the implications for professional practice.* Chichester, U.K.: Wiley.

Vullioud, C., Clément, F., Scott-Phillips, T. C., & Mercier, H. (2017). "Confidence as an expression of commitment: Why misplaced expressions of confidence backfire." *Evolution and Human Behavior*, 38(1), 9–17.

Walter, N., & Murphy, S. T. (2018). "How to unring the bell: A meta-analytic approach to correction of misinformation." *Communication Monographs*, 85(3), 1–19.

Wang, S. (1995). *Failure of charisma: The Cultural Revolution in Wuhan.* New York: Oxford University Press.

Ward, B. E. (1956). "Some observations on religious cults in Ashanti." *Africa*, 26(1), 47–61.

Warren, Z. J., & Power, S. A. (2015). "It is contagious: Rethinking a metaphor dialogically." *Culture and Psychology*, 21(3), 359–379.

Watts, D. J. (2011). *Everything is obvious*: Once you know the answer.* New York: Crown Business.

Weber, E. (2000). *Apocalypses: Prophecies, cults, and millennial beliefs through the ages.* Cambridge, MA: Harvard University Press.

Webster, S. W., & Abramowitz, A. I. (2017). "The ideological foundations of affective polarization in the US electorate." *American Politics Research*, 45(4), 621–647.

Wedeen, L. (2015). *Ambiguities of domination: Politics, rhetoric, and symbols in contemporary Syria.* Chicago: University of Chicago Press.

Weinberg, S. B., & Eich, R. K. (1978). "Fighting fire with fire: Establishment of a rumor control center." *Communication Quarterly*, 26(3), 26–31.

Weinberger, S. (2010). "Airport security: Intent to deceive?" *Nature*, 465(7297), 412–415.

Weisberg, D. S., Keil, F. C., Goodstein, J., Rawson, E., & Gray, J. R. (2008). "The seductive allure of neuroscience explanations." *Journal of Cognitive Neuroscience*, 20(3), 470–477.

Weisbuch, M., & Ambady, N. (2008). "Affective divergence: Automatic responses to others' emotions depend on group membership." *Journal of Personality and Social Psychology*, 95(5), 1063–1079.

Westfall, J., Van Boven, L., Chambers, J. R., & Judd, C. M. (2015). "Perceiving political polarization in the United States: Party identity strength and attitude extremity exacerbate the perceived partisan divide." *Perspectives on Psychological Science*, 10(2), 145–158.

Westwood, S. J., Peterson, E., & Lelkes, Y. (2018). *Are there still limits on partisan prejudice?* Working paper. Retrieved from https://www.dartmouth.edu/~seanjwestwood/papers/stillLimits.pdf

Whedbee, K. E. (2004). "Reclaiming rhetorical democracy: George Grote's defense of Gleon and the Athenian demagogues." *Rhetoric Society Quarterly*, 34(4), 71–95.

White, J. W. (2016). *Ikki: Social conflict and political protest in early modern Japan*. Ithaca, NY: Cornell University Press.

Williams, G. C. (1966). *Adaptation and natural selection*. Princeton, NJ: Princeton University Press.

Willis, R. G. (1970). "Instant millennium: The sociology of African witch-cleansing cults." In M. Douglas (Ed.), *Witchcraft confessions and accusations* (pp. 129–140). London: Routledge.

Winterling, A. (2011). *Caligula: A biography*. Los Angeles: University of California Press.

Wirtz, J. G., Sparks, J. V., & Zimbres, T. M. (2018). "The effect of exposure to sexual appeals in advertisements on memory, attitude, and purchase intention: A meta-analytic review." *International Journal of Advertising*, 37(2), 168–198.

Wiswede, D., Koranyi, N., Müller, F., Langner, O., & Rothermund, K. (2012). "Validating the truth of propositions: Behavioral and ERP indicators of truth evaluation processes." *Social Cognitive and Affective Neuroscience*, 8(6), 647–653.

Wlezien, C. (1995). "The public as thermostat: Dynamics of preferences for spending." *American Journal of Political Science*, 39(4), 981–1000.

Wohlstetter, R. (1962). *Pearl Harbor: Warning and decision*. Stanford, CA: Stanford University Press.

Wood, J., Glynn, D., Phillips, B., & Hauser, M. D. (2007). "The perception of rational, goal-directed action in nonhuman primates." *Science*, 317(5843), 1402–1405.

Wood, T., & Porter, E. (2016). *The elusive backfire effect: Mass attitudes' steadfast factual adherence*. Retrieved from https://papers.ssrn.com/sol3/papers.cfm?abstract_id=2819073

Wootton, D. (2006). *Bad medicine: Doctors doing harm since Hippocrates*. Oxford: Oxford University Press.

Wootton, D. (2015). *The invention of science: A new history of the scientific revolution*. London: Harper.

Wray, M. K., Klein, B. A., Mattila, H. R., & Seeley, T. D. (2008). "Honeybees do not reject dances for 'implausible' locations: Reconsidering the evidence for cognitive maps in insects." *Animal Behaviour*, 76(2), 261–269.

Wright, J. (1997). "Helping-at-the-nest in Arabian babblers: Signalling social status or sensible investment in chicks?" *Animal Behaviour*, 54(6), 1439–1448.

Wright, J., Parker, P. G., & Lundy, K. J. (1999). "Relatedness and chick-feeding effort in the cooperatively breeding Arabian babbler." *Animal Behaviour, 58*(4), 779–785.

Wright, R. (2009). *The evolution of God*. New York: Little, Brown.

Yamagishi, T. (2001). "Trust as a form of social intelligence." In K. Cook (Ed.), *Trust in society* (pp. 121–147). New York: Russell Sage Foundation.

Yang, J., Rojas, H., Wojcieszak, M., Aalberg, T., Coen, S., Curran, J., . . . Mazzoleni, G. (2016). "Why are 'others' so polarized? Perceived political polarization and media use in 10 countries." *Journal of Computer-Mediated Communication, 21*(5), 349–367.

Yaniv, I. (2004). "Receiving other people's advice: Influence and benefit." *Organizational Behavior and Human Decision Processes, 93*(1), 1–13.

Yaniv, I., & Kleinberger, E. (2000). "Advice taking in decision making: Egocentric discounting and reputation formation." *Organizational Behavior and Human Decision Processes, 83*(2), 260–281.

Zahavi, A., & Zahavi, A. (1997). *The handicap principle: A missing piece of Darwin's puzzle*. Oxford: Oxford University Press.

Zeifman, D. M., & Brown, S. A. (2011). "Age-related changes in the signal value of tears." *Evolutionary Psychology, 9*(3), 147470491100900300.

Zimbardo, P. G., Johnson, R., & McCann, V. (2012). *Psychology: Core concepts with DSM-5 update* (7th ed.). Boston: Pearson Education.

Zipperstein, S. J. (2018). *Pogrom: Kishinev and the tilt of history*. New York: Liveright.

Zollo, F., Bessi, A., Del Vicario, M., Scala, A., Caldarelli, G., Shekhtman, L., . . . Quattrociocchi, W. (2017). "Debunking in a world of tribes." *PloS One, 12*(7), e0181821.

INDEX

Page numbers in italics refer to figures.

absurd ideas, of scientists, 217–18
Acerbi, Alberto, 208, 299n36
action, false rumors and lack of, 153–54
Adaptation and Natural Selection (Williams), 22
adaptations, 22; in communication, 18
adaptive credulity, 10–13; anthropologists on, 9
adversarial relationships, communication and, 22–23
advertisers, xviii; celebrities and, 142–43; cost of, 141; negligible effects from, 141–42; political campaigns and, 141; preconceived opinions and, 141; television cigarette, 142; Tellis on, 143
Against Democracy (Brennan), 264
aggregation: majority opinion and, 71; Munroe *xkcd* "Bridge" comic strip on, 71, 72; Surowiecki on, 71
alarm calls, 21; of Arabian babbler, 22, 23; kin selection and, 22
alignment, of incentives, 84–85, 86, 88, 92, 282n24, 283n29
Allcott, Hunt, 213
Allport, Gordon, 147

analytic thinking, 45; Gervais and Norenzayan on, 37–38
animal behavior: of Arabian babbler, 16–17, 22, 23; of baboons, 71–72; of bees, 17–18, 19; of bowerbird, 16, 26–27; of chimpanzees, 40–41; pregnancy and, 17; of Thomson's gazelles, 16, 24–25, 28, 101; of vervet monkey, 18–19, 20, 40, 275n11
Anthony, Dick, 123–24
anti-Semitic propaganda, 128–29
anxiety, rumors and, 147–48
Arabian babbler, 16–17; alarm calls of, 22, 23
Arceneaux, Kevin, 137
Arendt, Hannah, 232
argumentation: common ground and, 62; counterintuitiveness and, 221–22; beyond plausibility checking, 50–55; small group discussion and, 113
arguments: challenging, 55–58; confidence in, 55; logical problems and, 51–52; reasoning in, 52–53; strength of, 56
arms race analogy, for open vigilance mechanisms, 31–32, 38, 41, 46

351

INDEX

Art of Deception, The (Mitnick), 249–50
Asch, Solomon, 5–6, 6, 74–75
automatic cognitive mechanisms, 100, 101–2, 105

baboons, majority opinion and, 71–72
backfire effect, 278n3; of Bush and Iraq War, 48–49; Nyhan and Reifler on, 48–49; vaccination opponents and, 49; Wood and Porter, E., on, 49
Bad Medicine (Wootton), 202
Bad Writing Contest, 218
Barker, Eileen, 122–23
Barrett, Justin, 222–23
Bataclan attacks, 111–12
Baumard, Nicolas, 229
bees, animal behavior of, 17–18, 19
beliefs: argumentation and plausibility checking, 50–55; causal effects, 214–16; challenging arguments and, 55–58; contrary opinions and, 48–50; costly actions and, 261; false rumors and, 151–55; intuition and, 58–62, 152; intuitive, 152, 178, 260, 261; justifications for, 214; misconceptions and, 260; preexisting, 47–48; reflective, 152, 178–79, 189–90, 196, 260–61; self-incriminating statements and, 197–98; social transmission of religious, 175, 177; Sperber on, 152
believers, 175–78
bias: frequency-based, 275n36; prestige, 12–13; success, 11–13
blind trust, 2–5
bloodletting practice: culture and, 203; Galen on, 199–200, 201, 207, 228–29
Bordia, Prashant, 149
bowerbird, 16; costly signaling of, 26–27
Boyd, Robert, 10, 275n32; on celebrity advertising, 142; on cultural learning and success, 11

Boyer, Pascal, 220; on information, 226–27
brainwashing, xviii; Gallery on, 33; McCarthyism, 32; open-mindedness and, 32–38, 42–46; of POWs, 32–33, 42–43
Brennan, Jason, 8, 264
Brexit, 35; fake news and, 200–201, 298n7
Broockman, David, 138, 140
Bryan, William Jennings, 116
Burgess, Thomas, 99
burning-bridges strategy, 192, 194; extreme beliefs defense in, 196–97; extreme flattery and, 191, 193; extreme views and, 195; intelligence or moral standing and, 195; reflective beliefs and, 196; self-incriminating statements in, 197–98
Burns, Justine, 254–55
Bush, George W., 212; backfire effect and, 48–49; Iraq War justification by, 172–74; 2000 presidential election, 137–38
Butler, Judith, 218

cable news networks, taking sides strategy of, 242–43
Cacioppo, John, 98
Cambridge Analytica, 139
campaigners, 134, 141, 290n53; ambiguous results of, 135–36; Arceneaux and Johnson on, 137; Cambridge Analytica and, 139; effectiveness experiments, 138; Gelman and King on, 140; inefficiency of, 139–40; Kalla and Broockman on, 138, 140; Klapper on, 136; lab-based techniques and, 136–37; media influence, 136–37, 140; 2000 presidential election and, 137–38; in U. S. politics, 135

INDEX

Canetti, Elias, 97
Caplow, Theodore, 149, 150
cascade, of influence, 285n56
Catholic Church: Children's Crusade and, 2; Enlightenment and, 264; mass persuasion and, 144; preachers and, 124–25, 127
celebrities: advertisers and, 143; Boyd and Richerson on advertising and, 142; prestige bias and suicide of, 12; relevant cultural products and, 156–57
challenging arguments, 55–58
charismatic authority, xiv; counterintuitiveness and, 225–26; of Lacan, 225
Chiarella, Sabrina, 103–4
children: culture continuity and, 9, 274n27; Dawkins on gullibility of, 9; gullibility of, 9, 45–46; incentives and, 86–87; intuition displayed by, 68–69; open vigilance mechanisms and, 248; selective ignorance of, 103
Children's Crusade, 2; Pope Innocent III influence on, 3
chimpanzees, communication signals of, 40–41
China Cultural Revolution, 132–34, 289n37
Chopra, Deepak, 238–39, 303n59
Christians: millenarian movements, 120–21; Stark on, 122
Cleon, 114–16
Clinton, Bill, 2
Clinton, Hillary, 2, 212, 260; fake news on, 201, 205
Clooney, George, 142–43
coarse cues, for trust, 240–41, 247–50, 254, 255
cognitive mechanisms: automatic and mandatory, 100, 101–2, 105; to find allies, 241; gullibility and, 257

cognitive sophistication: credulity and, 35; gullibility association with, 38; open-mindedness and, 38–42
commitment signals, 89; epistemic modals for, 90; Tenney on, 90–91
communication: adaptations in, 18; adversarial relationships and, 22–23; animal behavior and, 16–18; conflicts and evolution of, 18–20; cues in, 18–19; diligence in, 92; emotional signals in, 104–5; failures in, 20–22; omnivorous diets evolution analogy, 39–42; signals in, 18, 25–28; success in, 22–25; vigilance in, 15–29
Communist Party, Chinese, 133
Company of Strangers, The (Seabright), 240
competence: best knowledge and, 76–77; in performance, 68; preschoolers on, 76; in wide audience, 113
con men: 419 Nigerian scam, 250–51; in *The Sting*, 248–49; Thompson as, 249–50
Condorcet, Marquis de, 3, 71
Condorcet jury theorem, 71, 73
confessions: eyewitness testimony and, 182; interrogators and, 184, 295n14; in Japan, 185; Kassin and Wrightsman on, 184; shame and, 295n23; of witches, 185–90
conformity: Asch experiments on, 5–6, 6, 74–75; Gallup on, 75–76; Milgram experiments on, 6–8, 75, 232–33; Moscovici on, 5–6
conformity bias, 13; cultural learning and, 11; Japanese kamikaze and, 12
Conis, Elena, 60
conspiracy theories, 164, 172, 269–70; of Jones, 4, 228; as threat, 158

contagion analogy, 105, 108; on crowds, 96; Espinas on, 98; in New York, 95; pathogens and, 97, 106–7; Sighele on moral, 96; social media and, 96–97; in Tanganyika, 95–96; transmission of emotions and, 106
contrary opinions, beliefs and, 48–50
control, of facial expressions, 100, 284n26
Correa, Angela, 181
costly actions, beliefs and, 261
costly signals: bowerbirds and, 26–27; in communication, 25–26, 241–42; Zahavi on, 26
counterempathy, 105
counterintuitive scientific theory, 231–33, 237–38, 270
counterintuitiveness, 218; argumentation and, 221–22; charismatic authority, 225–26; concepts and, 219–20; on inertia, 224, 224; in intuition, 222–23; intuitive thinking and, 222–23; reflective beliefs and, 261; religious concepts and, 220, 222–23; scientific concepts and, 220–21, 223, 224; shallowness and, 225
credulity: adaptive, 9–13; cognitive sophistication and, 35; Gilbert experiments on, 36–37, 43–44; gullibility compared to, 273n4; Heraclitus on, 8–9; observers on, 4
crisis, rumors of, 147–48, 158–59
Crowd in the French Revolution (Rudé), 108–9
crowd psychology: Heraclitus on, 34; Le Bon, Tarde and Taine on, 34, 96; politics and, 34
crowds: contagion of feelings in, 96; panic in, 111–12; rational, 108–12
Crucible, The (Miller), 185

crusades, 2, 3, 126
cues, 161; for changing mind, 259; coarse, for trust, 240–41, 247–50, 254, 255; in communication, 18–19; evolutionarily valid, 73, 74; liars nonverbal, 78–79
cultural learning: conformity bias and, 11; success bias and, 11; from successful individuals, 11
culture: bloodletting practice and, 203; children and continuity of, 9, 274n27; exhaustive cultural transmission and, 9; human survival and, 10; maladaptive practices in, xiv, 13; religious beliefs and, 294n21
Cunningham, Steven, 181
curiosity, about rumors, 155–59

Dalai Lama, 294n15
Dao, David, 146, 165, 292n2
Darjeeling landslide, Sinha on rumor of, 147
Darwin, Charles, 99
Dawkins, Richard, 9
death penalty, justifications for, 210
deception detection, 78
demagogues: of Bryan and Long, 116; Cleon, 114–16; existing opinions relied on by, 114–18; of Hitler, 116–21
democracy, 210, 211, 264; Plato on, 3
Democrats, MSNBC and liberal, 242–43
Deskovic, Jeffrey, 181, 183
Dezecache, Guillaume, 98, 102
DiFonzo, Nicholas, 149
Diggory, James, 150–51
diligence, 83–84, 92, 282n20
Dimberg, Ulf, 97–98
discussion groups, polarization in, 209

INDEX 355

Dockendorff, Martin, 71
Duna, 294n16; reflective beliefs of, 178; religious beliefs of, 176–77; San Roque on, 176
Echols, Catharine, 86–87
economic games experiments, on trust, 254–55, 304n28
Eich, Ritch, 147
Ekman, Paul, 79–80
Emotional Contagion (Hatfield, Cacioppo, Rapson), 98
emotional signals, in communication, 104–5
emotional vigilance, 104; automatic and mandatory mechanisms in, 101–2, 105; children selective ignorance and, 103
Engels, Friedrich, 13, 124
Englis, Basil, 97, 105
Enigma of Reason, The (Sperber), 57
Enlightenment, 3, 126, 263; Catholic Church and, 264
entertainment, Acerbi on fake news, 208, 299n36
epistemic modals, for commitment, 89–90
epistemic vigilance, Sperber on, 31
Eriksson, Kimmo, 231–32
Espinas, Alfred, 98
Eusebius, 122
Evans-Pritchard, E. E., 186
evidentials: of Duna, 176–77; of Kaluli language, 178–79; in language, 168–69
evolution: of communication, 18–20; by natural selection, 19; of omnivorous diets analogy, 39–42
evolutionarily valid cues, 73, 74
exonerations, in false confessions, 182

extreme beliefs defense, in burning bridges strategy, 196–97
extreme flattery, in burning bridges strategy, 191, 193
extreme views, in burning bridges strategy, 195
eyewitness advantage, 65; informational access and, 64

face recognition, 156
facial expressions, 79, 80, 98–99; control of, 100, 284n26
failures, in communication, 20–22
fake news, 199; Acerbi on entertainment of, 208, 299n36; beliefs causal effects and, 214–16; of Brexit, 200–201, 298n7; on Clinton, H., 201, 205; *Collins* dictionary on, 200; justifications and, 206–8; polarization and, 208–11; political, 207–8; sensationalism and, 215–16; in social media, 207, 298n11, 299n32; Trump election and, 200–201, 204–5, 207, 215, 298n7; U.S. polarization, 211–14
false beliefs, 202, 266; from trust, 245
false confessions, 197; coerced, 182; of Deskovic, 181, 183; exonerations and, 182; persuasion and, 182; voluntary, 182
false rumors, 148–50, 263, 269, 292n13; belief in, 151–55; lack of action following, 153–54; social costs of, 161–62, 171–72; about threats, 157–58; on Twitter, 158
fax model of cultural transmission, 9, 10
feelings: of anger, 100; contagion of, 95–98, 105–8; Darwin on, 99; emotional vigilance, 101–5; expression of, 100; Frank on, 99–100; passion without reason, 98–101; pathogens and, 97, 106–7; rational crowds and, 108–12

Fershtman, Chaim, 254
Fiorina, Morris, 211
flattery, extreme, 191, 193, 296n40
flattery inflation, Márquez on, 195
419 Nigerian scam, 250–51
Fox News Channel: conservative Republicans and, 242–43, 245–46; studies on politics effects of, 245–46
Frank, Robert, 99–100
French Revolution, 108–9
frequency-based bias, 275n36
Freud, Sigmund, 78
friction and flooding, Roberts on, 133

Galen: bloodletting practice by, 199–200, 201, 207, 228–29; humoral theory of disease support by, 199–200
Gallery, Daniel, 33
Gallup, Andrew, 75–76
Galton, Francis, 71
Gardner, Howard, 68
gazelles. *See* Thomson's gazelles
Gelman, Andrew, 140, 246
Gendelman, Max, 86
gene-culture coevolution, 13; Boyd and Richerson on, 10
Gentzkow, Matthew, 212–13
German soldiers studies, and Nazi propaganda, 130–31
Germinal (Zola), 262–63
Gervais, Will, 37–38
Gilbert, Daniel: credulity experiments on, 36–37, 43–44; on gullibility, 8
Gil-White, Francisco: on prestige bias, 12; on success bias, 11–12
Gneezy, Uri, 254
God: Barrett on canonical features of, 223; omniscience of, 217, 223, 230, 302n34; religious concept of, 220
Gödel, Kurt, 56

Goebbels, Joseph, 128, 129, 135
Goldman, Alvin, 238
Gore, Al, 33, 212; 2000 presidential election, 137–38
groups: affiliation signals for, 241; argumentation and small group, 113; membership costs in, 191–92; polarization in discussion, 209
gullibility, 13–14; of children, 9, 45–46; cognitive sophistication association with, 38; credulity compared to, 273n4; Dawkins on children, 9; examples of, xiii–xiv; Gilbert on, 8; gullible about, 262–65; Trump election and, 35, 276n15
guru effect, 238; Lacan and, 234–36; obscure statements and, 234; Sperber on, 234

Haig, David, 21
Handbook of the Law of Evidence (McCormick), 182
Hatfield, Elaine, 98
Henrich, Joe: on cultural learning and success, 11; on prestige bias, 12; on success bias, 11–12
Hepach, Robert, 104
Heraclitus: on credulity, 8–9; on crowd psychology, 34
hidden dependencies: open vigilance mechanisms and, 174; religious beliefs and, 176; sourcing and, 172–75
historical evidence, 56–57
History of the Peloponnesian War (Thucydides), 167
Hitler, Adolf, 34; as demagogue, 116–21; as propagandist, 128; Selb and Munzert on, 116
Hitler Myth, The (Kershaw), 116
Ho Chi Minh, 191

INDEX 357

humoral theory of disease, 202–4, 214–15; Galen support of, 199–200
Hussein, Saddam, 48–49, 173, 190

Iannaccone, Laurence, 124
Icke, David, 172
ideological polarization, 212–13
illusion of unanimity, 112
imitation, reliable expertise and, 66–67
immigration, Trump on, 268
incentives, 282n26; alignment of, 84–85, 86, 88, 92, 282n24, 283n29; children and, 86–87; Gendelman and Kirschner example, 86; reputation monitoring, 88; Reyes-Jaquez and Echols experiment on, 86–87; Sniezek study on, 85–86; social alignment of, 89; trust and, 84–87
inclusive fitness, 19, 20
individuals: cultural learning from successful, 11; signals for affiliation of, 241; stock in majority opinion, 70–71
inefficiency, of campaigners, 139–40
inferences, 59, 170
influence: cascade of, 285n56; difficulty of, xvi; power of, 15
information: Boyer and Parren on, 226–27; gains, trust and, 252–54; rejection, 93; relevance of, 159–60; social cost of inaccurate, 246; social relevance of, 159–60; spread of, 160, 161
informational access: experiments on, 64, 65; eyewitness advantage and, 64; Robinson on, 64
informational environment, open vigilance mechanisms and, xvii
InfoWars website, 228
Innocent III (pope), Children's Crusade influenced by, 3
insight problems, 51

intelligence, 68, 195
interlocutors: opinion of, 283n31; sourcing and, 166, 170–71
interrogators, confessions and, 184, 295n14
intuition, xvii, 277n22; beliefs and, 58–62, 152; children display of, 68–69; counterintuitiveness and thinking in, 222–23; soundness of, 54
intuitive beliefs, 152, 261; misconceptions of, 260; religious beliefs and, 178
intuitive physics, 223–24, 224
Iraq War: backfire effect and, 48–49; Bush justification for, 172–74

Japan: confessions in, 185; kamikaze, conformity bias of, 12
Johnson, Martin, 137
Jones, Alex: conspiracy theories of, 4, 228; Welch influenced by, 4
Jordan, Michael, 142
justifications, 237; for alternative treatments, 206–7, 214; for beliefs, 214; competition and, 207; for death penalty, 210; fake news and, 206–8; of Iraq War, by Bush, 172–74; for negative judgments, 206; polarization and, 208–9; reputation credit and, 227–29

Kahneman, Daniel, 35–36, 37
Kalla, Joshua, 138, 140
Kaluli language, evidentials of, 178–79
Kassin, Saul, 184
Kay, Jonathan, 153
Kershaw, Ian, 116, 118; on Nazi propaganda, 129–30, 131, 259
Kierkegaard, Søren, 70
Kim, Eunji, 205
Kim, Jin Woo, 205

Kim, Young Oon, 123
Kim Jong-il, xiv, xviii, 190–91, 195;
    extreme flattery and, 193, 296n40
Kim Jong-un, 190, 296n40
kin selection, alarm calls and, 22
King, Gary, 140, 246
Kirschner, Karl, 86
Kishinev accusations, rumors on, 200, 204, 215
Klapper, Joseph, 136
knowledge, best: competence and, 76–77; eyewitness advantage, 64–65; majority pull and, 74–76; past performance, 67–69; rationality, 70–74; reliable expertise, 65–67
Koji, Aoki, 110
Korean War, POWs brainwashing in, 32–33, 42–43

Lacan, Jacques, 218, 238, 239; charismatic authority of, 225; guru effect and, 234–36; obscurity of, 234–36; teachings of, 219
Laland, Kevin, 11
language: and evidentials, 168–69; and Kaluli evidentials, 178–79; Wanka Quechua, on sourcing, 168–69
Lanzetta, John, 97, 105
Latour, Bruno, 219
leaders, charismatic, xiv, 225–26
learning: cultural, 11; open vigilance mechanisms for, 258–59; transfer effects in, 280n11
Le Bon, Gustave, 34, 96
Le Roy Ladurie, Emmanuel, 125
Levine, Tim, 82
Lévi-Strauss, Claude, 225, 236
liars, 282n17; detection of, 78; Levine on, 82; nonverbal cues of, 78–79; Reid on, 82

Lie to Me television show, 79
logical problems, arguments and, 51–52
Long, Huey, 116
Luther, Martin, 58
lying, trust and, 81–82
Lysenko, Trofim, 266

Madden, Joah, 27
majority opinion, xiv; assembly example, 70–71, 72, 73; of baboons, 71–72; Dockendorff, Schwartzberg, Mercier on, 71; evolutionarily valid cues and, 73, 74; Galton on aggregation and, 71; individuals stock in, 70–71; Kierkegaard on, 70; Morgan experiment on, 72–73; Munroe *xkcd* "Bridge" comic strip on, 71, 72; resistance to, 74–76; Twain on, 70
maladaptive cultural practices, xiv, 13
mandatory cognitive mechanisms, 100, 101–2, 105
Mao Zedong, 266–67
Maps of Meaning (Peterson), 238
Márquez, Xavier, 133; on flattery inflation, 195
Marx, Karl, 13, 124
mass conversions, from preachers, 122
mass persuasion, xviii, 14, 259; Catholic Church and, 144; patterns of, 143; plausibility checking and, 113–14; resistance and, 144
mass psychogenic illness, 106–8
McCain, John, 205
McCarthy, Jenny, 60
McCarthyism, 32
McCloskey, Michael, 223
McCormick, Charles, 182
media: campaigners influence and, 136–37, 140; Gelman and King on

INDEX 359

politics influenced by, 246; mass, 160. *See also* fake news; social media
Mein Kampf (Hitler), 128
membership costs, in groups, 191–92
Meno (Plato), 53
Mercier, Hugo, 71, 73
microexpressions, trust and: Ekman on, 79–80; Porter, and ten Brinke on, 80–81
Milgram, Stanley, 6–8, 75, 232–33
millenarian movements: Christian, 120–21; prophets and, 119–20; Weber on, 120
Miller, Arthur, 185
minimal plausibility, rumors and, 160–62
misperception, of partisanship and polarization, 244
Mitnick, Kevin, 249–50
Miton, Helena, 74
Moonies, 121, 123; Barker on, 122
moral contagion, 96
Moreau, Sabine, 63, 64
Morgan, Thomas, 72–73
Morin, Edgard, 167
Mormonism, 121, 122
Moscovici, Serge, 5–6
movements: Cattle-Killing, 118–19; millenarian, 119–21; New Religious Movements, 121–24; in public opinion, 268; Truth, 153
MSNBC, liberal Democrats and, 242–43
Munroe, Randall, 71
Munzert, Simon, 116
Mussolini, Benito, 34

natural selection, 28; evolution by, 19
Nazi propaganda, 143–44; Kershaw on effectiveness of, 129–30, 131, 259

negligence, 83–84
New Religious Movements, 121, 122; Anthony on, 123–24
9/11 terrorist attacks: rational crowds and, 111–12; reflective beliefs on, 260; rumors of, 165
Nongqawuse, Xhosa influenced by, 2–4, 118
nonverbal cues: Freud on, 78; for liars, 78–79
Norenzayan, Ara, 37–38
Nyhan, Brendan, 48–49, 205

Obama, Barack, 146, 205–6
omniscient God, religious beliefs of, 217, 223, 230, 302n34
omnivorous diets evolution analogy, 39–42
open vigilance mechanisms, xv, 292n20; arms race analogy for, 31–32, 38, 41, 46; burning bridges and, 191–98; children and, 248; confessions and, 182–90, 295n23; cues for, 18–19, 73–74, 78–79, 161, 240–41, 247–50, 255; current informational environment and, xvii; hidden dependencies and, 174; information rejection and, 93; for learning, 258–59; mass persuasion and, xviii, 14, 113–14, 133, 143–44, 259; motivations for, xviii; open-mindedness and, 30–46, 54, 58, 63; plausibility checking in, 47–48, 50–55, 113–14, 221; psychological experiments on, 144–45; reasoning and, 52–54, 58, 98–101; sourcing and, 166–75, 238n38
open-mindedness, 30–31, 63; cognitive sophistication and, 38–42; reasoning and, 54, 58

opinion: advertisers and preconceived, 141; beliefs and contrary, 48–50; convergence of, 174; demagogues' reliance of existing, 114–18; of interlocutors, 283n31. *See also* majority opinion; public opinion
Origgi, Gloria, 179
Osborne, Sarah, 185–86
Osnos, Evan, 132
overconfidence: reputation and, 91–93; trust and, 90–92, 283n37

panic, in crowds, 111–12
pareidolia, 157
Parren, Nora, 226–27
partisanship: of cable news networks, 242; misperception of, 244
Passions within Reason (Frank), 99
past performance: best knowledge and, 67–69; evaluation of, 66; reliable expertise and, 65–66; reputation credit and, 226
pathogens: Canetti on, 97; contagion analogy and, 97, 106–7
Peires, Jeff, 118
Peloponnesian War, 1
performance: examples of, 69; observations of, 68; from observed to competence, 68; past, 65–69, 226
persuasion: false confessions and, 182. *See also* mass persuasion
Peterson, Jordan, 238
Planck, Max, 56
Plato, on democracy, 3
plausibility checking, 47–48, 221; argumentation beyond, 50–55; insight problems and, 51; mass persuasion and, 113–14
polarization: on death penalty, 210; in discussion groups, 209; fake news and, 208–11; Gentzkow and Shapiro on ideological, 212–13; justifications and, 208–9; misperception of, 244; political, 210, 213; social media and, 210–11 polarization, U.S., 214; Fiorina on, 211; impression of increased, 212; in politics, 211; social media users and, 212–13, 244
politics: advertisers and campaigns in, 141; crowd psychology and, 34; fake news in, 207–8; Gelman and King on media influence on, 246; polarization in, 210, 213; public opinion and, 267–68; trust and, 94; U.S. polarization in, 211
Porter, Ethan, 49
Porter, Stephen, 80–81
Postman, Leo, 147
Poulin-Dubois, Diane, 103–4
Pound, John, 149
Powell, Colin, 173
preachers: Catholic Church and, 124–25, 127; crusades and, 2, 3, 126; Eusebius on, 122; mass conversions and, 122; Mormonism and, 121, 122; New Religious Movements, 121, 122, 123–24
predator-deterrent signals, 23–24
preexisting beliefs, 47–48
pregnancy, 17, 20, 21, 28
prestige bias, 13; celebrity suicide and, 12; Henrich and Gil-White on, 12
prisoners of war (POWs), 32–33, 42–43
propagandists, 264–65; China Cultural Revolution, 132–34, 289n37; failures of, 133; Goebbels, 128, 129; Hitler as, 128; Soviet, 131–32; threats and, 134
prophets, 117, 121; millenarian movements of, 119–20; Xhosa and Nongqawuse as, 2–4, 118–19

INDEX 361

Psychology of Rumor, The (Allport and Postman), 147
public opinion: movements in, 268; politics and, 267–68; thermostatic model of, 268
punishment, for unreliable messages, 88
Putin, Vladimir, 265, 305n18; Trump and, 267

Rapson, Richard, 98
rational crowds: Koji on, 110; Bataclan attacks and, 111–12; England peasant revolt and, 109–10; in French Revolution, 108–9; illusion of unanimity and, 112; 9/11 terrorist attacks and, 111–12; panic and, 111–12; Red Guards spontaneous mobs, 110; Shays' Rebellion, 110; soldiers and, 112
reasoning: in arguments, 52–53; open-mindedness, 54, 58; vigilance and, 54
reflective beliefs, 152; burning bridges strategy and, 196; counterintuitiveness and, 261; of Duna, 178; on 9/11 terrorist attacks, 260; Origgi on, 179; religious beliefs and, 178; in witchcraft, 189–90
Reid, Thomas, 82
Reifler, Jason, 48–49
reliable expertise: best knowledge and, 65–67; imitation and, 66–67; past performance and, 65–66; preschoolers on, 76
religious beliefs, 288n68, 288n75; Baumard on, 229; culture and, 294n21; of Duna, 176–77; hidden dependencies and, 176; of omniscient God, 217, 223, 230, 302n34; reflective instead of intuitive, 178; social transmission of, 175, 177; variety of, 217; in world religions, 230

religious concepts: counterintuitiveness and, 220, 222–23; of God, 220
religious people, trust in, 247
#Republic: Divided Democracy in the Age of Social Media (Sunstein), 210
Republicans, Fox News Channel and conservative, 242–43, 245–46
reputation: incentives and monitoring of, 88; overconfidence and, 91–93; trust and, 87–90
reputation credit, 230, 237; justifications and, 227–29; past performance and, 226; threats in, 226–27, 228; valuable information and, 226
Reyes-Jaquez, Bolivar, 86–87
Rice, Condoleezza, 173
Richerson, Peter, 10, 275n32; on celebrity advertising, 142; on cultural learning and success, 11 rituals, 10
Roberts, Margaret, 133
Robinson, Elizabeth, 64
Romans, Humbert de, 125–27
Rothbard, Murray, 194
Rudé, George, 108–9
rumeur d'Orléans, 146, 148, 153–55, 161–64, 200; Morin on, 167; sources and, 166, 171
rumors: acting on, 165; anxiety and, 147–48; of crisis, 147–48, 158–59; about Dao and United Airlines, 146, 165, 292n2; on Darjeeling landslide, 147; escape from reality and, 162–64; exaggerated threats and, 261–62; on Kishinev accusations, 200, 204, 215; metarumors and, 295n40; minimal plausibility and, 160–62; of 9/11 terrorist attacks, 165; about Obama, 146, 205–6; rabies outbreak, 150–51; rewarding relays of, 159–60; of

rumors (cont.)
   rumeur d'Orléans, 146, 148, 153–55, 161–64, 166–67, 171, 200; sourcing quality and, 166–67; spontaneous tracking of, 150–51; unfettered curiosity and, 155–59; about University of Michigan strike, 147, 151; wartime, 149; workplace, DiFonzo and Bordia on, 149; on World War II Japanese treason, 151, 153

Salem witch trials: Osborne and, 185–86; Tituba's confessions in, 186, 189
San Roque, Lila, 176
Schieffelin, Bambi, 179
Schwartzberg, Melissa, 71
scientific concepts, counterintuitiveness and, 220–21, 223, 224
scientific theory, counterintuitive, 231–33, 237–38, 270
scientists, absurd ideas of, 217–18
Scott, James, 127
Scott-Phillips, Thom, 102
Seabright, Paul, 240
Selb, Peter, 116
selective ignorance, of children, 103
self-deception, 280n18
self-incriminating statements, 197–98
sensationalism, fake news and, 215–16
Shapiro, Jesse, 212–13
Shays' Rebellion, 110
Sighele, Scipio, 96
signals, in communication, 18, 27–28; automatic emotional reactions, 97–98; of chimpanzees, 40–41; commitment, 89–90; costly, 25–26, 241–42; for individual or group affiliation, 241; predator deterrent, 23–24; unreliable, 20, 25–26, 88, 104; of vervet monkeys, 40

Signer, Michael, 114
Sinha, Durganand, 147
Sniezek, Janet, 85–86
social alignment, of incentives, 89
social cost: of false rumors, 161–62, 171–72; of inaccurate information, 246
social media: Allcott on political polarization and, 213; contagion analogy and, 96–97; fake news in, 207, 298n11, 299n32; polarization in, 210–11; U.S. polarization and users of, 212–13, 244
social relevance, of information, 159–60
social transmission, of religious beliefs, 175, 177
Socratic questioning, 53
sourcing: hidden dependencies, 172–75; interlocutors and, 166, 170–71; rumors and quality of, 166–67; trust and, 283n38; two degrees of separation, 171–72. See also evidentials
Soviet propaganda, 131–32, 143
Sperber, Dan, 31, 57; on beliefs, 152; on face recognition, 156; on guru effect, 234
spontaneous tracking, of rumors, 150–51
Stalin, Joseph, 32
Stark, Rodney, 122
Sternberg, Robert, 68
Stimson, James, 267
Sting, The film, con men in, 248–49
stotting, of Thomson's gazelles, 24–25, 28, 101
Strandburg-Peshkin, Ariana, 71–72
strangers, trust in, 240, 241, 247–51
subliminal influence, xviii, 43, 263, 278n46; brainwashing and, 33–34

INDEX   363

success bias: athletic products and, 12; Henrich and Gil-White on, 11–12; Marx and Engels on, 13
suggestibility, cost of, xv
Sun Myung Moon, 122
Sunstein, Cass, 210
Surowiecki, James, 71
sycophant, credible, 190–91
System 1 and System 2 thought processes, 35–38, 44–45, 277n20

Taine, Hippolyte, 34
taking sides strategy: of cable news networks, 242–43; to gain audiences, 243; with minimal costs, 242; misrepresentations spread by, 243
Tamis-LeMonda, Catherine, 103
Tanganyika, 107; contagion of feelings in, 95–96
Tarde, Gabriel, 34, 96
Tellis, Gerard, 143
ten Brinke, Leanne, 80–81
Tenney, Elizabeth, 90–91
thermostatic model, of public opinion, 268
Thinking, Fast and Slow (Kahneman), 35–36
Thompson, Samuel, 249–50
Thomson's gazelles, 16; stotting of, 24–25, 28, 101
threats, 237; conspiracy theories as, 158; false rumors about, 157–58; propagandists and, 134; in reputation credit, 226–27, 228; rumors about exaggerated, 261–62
Thucydides, 3, 167
Tituba, Salem witch trials confession by, 186, 189
transfer effect, in learning, 280n11

Trump, Donald, 212; fake news and election of, 200–201, 204–5, 207, 215, 298n7; gullibility and election of, 35, 276n15; on immigration, 268; Putin and, 267
trust, 78, 93–94, 222; blind, 2–5; calibration of, 255–57; coarse cues for, 240–41, 247–50, 254, 255; damage from breakdown in, 94; economic games experiments, 254–55, 304n28; effective irrational, 251–55; fragile chains of, 269–71; incentives and, 84–87; information gains and, 252–54; lying and, 81–82; negligence and, 83–84; overconfidence and, 90–92, 283n37; reputation and, 87–90; sourcing and, 283n38; in strangers, 240, 241, 247–51; taking sides strategy and, 244–45; in the wrong people, 240–41
Truth movement, Kay on, 153
Twain, Mark, 34–35; on majority opinion, 70
Twitter, false rumors and, 158
two degrees of separation, 171–72
2000 presidential election, of Bush and Gore, 137–38

Unification Church, Stark and Lofland on, 123
United Airlines, Dao and, 146, 165, 292n2
United States, polarization in, 211–14
University of Michigan strike rumor, 147, 151
unreliable signals, 20, 25–26, 88, 104

vaccinations, 61, 269; backfire effect and opponents of, 49; Conis on, 60; Wakefield and McCarthy on, 60
variety, of religious beliefs, 217
vervet monkey, 18–19, 20, 40, 275n11

Veyne, Paul, 265
vigilance: in communication, 15–29; epistemic, Sperber on, 31; need for, 28–29; in omnivorous diets, 39–40; reasoning and, 54
viral marketing, 96
Voigtländer, Nico, 128–29
Voltaire, 202
voluntary false confessions, 182
Voth, Hans-Joachim, 128–29

Wakefield, Andrew, 60
Wang, Shaoguang, 132
Wanka Quechua language, on sourcing, 168–69
"War of the Worlds" broadcast, of Welles, 96
wartime rumors, 149
Weber, Eugen, 120
Weinberg, Sandord, 147
Weisberg, Deena, 232
Welch, Edgar Maddison, 2, 4, 154–55
Welles, Orson, 96
Williams, George, 22
Wisdom of Crowds, The (Surowieski), 71

witchcraft: confessions of, 185–90; Evans-Pritchard study on, 186; reflective beliefs for, 189–90. *See also* Salem witch trials
Wood, Thomas, 49
Woods, Tiger, 94
Wootton, David, 202
workplace rumors, 149
world religions, 230
World War II, Japanese treason rumor, 151, 153
Wray, Margaret, 17–18
Wrightsman, Lawrence, 184

Xhosa: Cattle-Killing movement, 118–19; ghost army and, 2; Nongqawuse influence on, 2–4, 118
xkcd "Bridge" comic strip, of Munroe, 71, 72

Yamagishi, Toshio, 252–53

Zahavi, Amotz, 26
Zeckhauser, Richard, 149
Zola, Émile, 262–63